National Governance
and the Global Climate
Change Regime

National Governance and the Global Climate Change Regime

Dana R. Fisher

ROWMAN & LITTLEFIELD PUBLISHERS, INC.
Lanham • Boulder • New York • Toronto • Oxford

ROWMAN & LITTLEFIELD PUBLISHERS, INC.

Published in the United States of America
by Rowman & Littlefield Publishers, Inc.
A wholly owned subsidiary of The Rowman & Littlefield Publishing Group, Inc.
4501 Forbes Boulevard, Suite 200, Lanham, MD 20706
www.rowmanlittlefield.com

P.O. Box 317, Oxford OX2 9RU, UK

British Library Cataloguing in Publication Information Available

Library of Congress Cataloging-in-Publication Data

Fisher, Dana, 1971–
  National governance and the global climate change regime / Dana
Fisher.
    p. cm.
  Includes bibliographical references and index.
  ISBN 0-7425-3052-3 (alk. paper) — ISBN 0-7425-3053-1 (pbk. : alk.
paper)
  1. Climatic changes—Government policy. 2. Climatic changes—Economic
aspects. 3. Climatic changes—International cooperation. I. Title.
  QC981.8.C5F43 2004
  363.738'747—dc22
                              2004000311

Printed in the United States of America

♾ ™ The paper used in this publication meets the minimum requirements of
American National Standard for Information Sciences—Permanence of Paper
for Printed Library Materials, ANSI/NISO Z39.48-1992.

In memory of Morita-Sensei (1950–2003)

Contents

Figures and Tables

FIGURES

TABLES

Abbreviations

BP	British Petroleum
BTU	British thermal unit
COOL	Climate Options for the Long Term
COP-1	Conference of the Parties-1 (Berlin, 1995)
COP-2	Conference of the Parties-2 (Geneva, 1996)
COP-3	Conference of the Parties-3 (Japan, 1997)
COP-4	Conference of the Parties-4 (Buenos Aires, 1998)
COP-5	Conference of the Parties-5 (Bonn, 1999)
COP-6	Conference of the Parties-6 (The Hague, 2000)
COP-6bis	Conference of the Parties-6bis (Bonn, 2001)
EPA	Environmental Protection Agency
EU	European Union
GHG	greenhouse gas
GDP	gross domestic product
HEP	human exceptionalist paradigm
HFCs	hydrofluorocarbons
IEA	International Energy Agency
IPCC	Intergovernmental Panel on Climate Change
KNMI	Royal Netherlands Meterological Institute
LDP	Liberal Democratic Party
MITI	Ministry of International Trade and Industry
MTOE	million tonnes of oil equivalent

NEP	new ecological paradigm
NEPPs	national environmental policy plans
NGOs	nongovernmental organizations
OECD	Organisation for Economic Co-operation and Development
OLS	ordinary least squares
PFCs	perfluorocarbons
RIVM	Dutch National Institute of Public Health and the Environment
SMOs	social movement organizations
TPES	total primary energy supply
UNEP	United Nations Environment Programme
UNFCCC	United Nations Framework Convention on Climate Change
VNO-NCW	Confederation of Netherlands Industry and Employers
WMO	World Meteorological Organization

Acknowledgments

So many people have contributed to the success of this book that I am unable to mention them all here. First, I would like to thank my dissertation committee, at the University of Wisconsin, Madison, for their support and assistance: William R. Freudenburg, Frederick H. Buttel, Mustafa Emirbayer, Lewis Friedland, Pamela Oliver, and John Magnuson. In particular, Bill and Lew endured hundreds of hours talking me through the conception, research, and writing process. Bill and Lew continued to be great advisors as I revised the dissertation into a book manuscript.

In addition, I would like to show my gratitude to two particular people who made this research possible. First and foremost, Tsuneyuki Morita, who helped me to figure out what I really wanted to study. Through his assistance, I was able to meet so many wonderful people at his offices at the Japanese National Institute for Environmental Studies and at the Tokyo Institute of Technology. Second, I would like to thank Arthur Mol for providing support, advice, and office space while I was conducting field research in the Netherlands.

This project would not have been possible without the support of various funders: the National Science Foundation Integrative Graduate Education and Research Traineeship (IGERT) Program, the National Science Foundation Summer Institute in Japan, the National Science Foundation Dissertation Grant, and the Wageningen University Research Fellowship. Additionally, I would like to thank Jessica Gribble at Rowman & Littlefield.

Without my friends and family to support me, this book would never have come together: Among those whom I wish to express my sincere gratitude are Susan Mannon, for making me laugh when I was living out of a suitcase; Brack Hale, for letting me put my desk in the living room; Jessica Green and for her editing assistance; Doug "Slurp" Miller for his wonderful cover art; and Dad, Mom, Lauren, Carley, and my Memom, who asked questions and showed support throughout.

Finally, I would like to thank my partner, Aaron Patton, for living with me through the stress and long hours while I finished this project—you continue to be a great surprise to me every day.

1

Explaining the Regulation of the Global Environment: Theoretical Perspectives and Alternative Theories

In July 2001, the future of the international agreement to mitigate global warming—the Kyoto Protocol—was uncertain: President George W. Bush had announced that the United States would not be a party to the ratification of the treaty; Prime Minister Junichiro Koizumi had suggested that without the United States, the Japanese government might not consider ratification; and without Japanese ratification, the protocol might not have enough countries signed on to account for the requisite 55 percent of the Annex I countries' carbon dioxide emissions that was needed to bring the climate change treaty into legal force. On the third day of the second part of the Conference of the Parties-6 (COP-6bis) climate change negotiations in Bonn, Germany, Japanese environment minister Yoriko Kawaguchi stated:

> Japan will exert its utmost efforts to make it possible for many countries, including Japan, to conclude the Protocol. Aiming at the entry into force of the Kyoto Protocol by 2002. . . . In order to pursue effective measures against global warming, it is important that all countries act under one single rule. . . . To have the U.S. participation for the early entry into force of the Kyoto Protocol is by far the best scenario. (Kawaguchi 2001a)

This statement did nothing to assuage fears that, after almost ten years of negotiations leading up to a legally binding agreement, the Kyoto Protocol would fail to become an environmental treaty and all of the work would have been in vain. Even high-ranking members of international organizations such as the Intergovernmental Panel on Climate Change (IPCC) expected that, in the words of Taka Hiraishi, the cochair of the IPCC Task Force on National Greenhouse Gas Inventories, the meeting in Bonn would mark the end of the Kyoto Protocol—and the next round of negotiations, scheduled for fall 2001 in Marrakech, Morocco, would serve as a "postmortem" (interview, Hiraishi, July 18, 2001). The European Union (EU), however, was on a "rescue operation of the Kyoto Protocol," as EU delegation spokesperson Margot Wallerström said: "We do not have any alternatives" (Wallerström 2001). High-level negotiating teams representing all major economic powers except the United States plunged into negotiations to try to come up with an acceptable final text regarding the mechanisms through which the emission reductions stipulated in the Kyoto Protocol would be met (for a full discussion of the Kyoto mechanisms under discussion in the COP-6 round of negotiations, see Kopp 2001; see also Müller 2001). The day after these high-level meetings were scheduled to end, and after almost forty-eight hours of nonstop negotiations, representatives from 178 nations around the world agreed to the Bonn Agreement, "designed to finalise the text of the Kyoto Protocol and to strengthen the implementation of the UN Framework Convention on Climate Change" (Müller 2001, 1).

Although social movement organizations and representatives of some nations criticized the final agreement for being, in the words of a Greenpeace press statement, "Kyoto-Lite," many members of delegations and organizations involved in the negotiations applauded the compromise that made the agreement possible. Even as countries were discussing plans to move forward and ratify the protocol, thereby bringing into legal force the international treaty to mitigate global climate change, Paula Dobriansky, the U.S. undersecretary of state for global affairs, continued to push the American position against the protocol. During the final plenary session of the ministerial meeting in Bonn, Dobriansky stated that "the United States must emphasize that our not blocking consensus on the adoption of these Kyoto Protocol rules does not change our view that the Protocol is not sound policy" (2001, 5). With the Kyoto Protocol moving into the domestic ratification stage, all advanced nations except for the United States continued to be parties to it, preparing for the treaty's ratification while developing state and market tools to restrain national emissions.

In this book, I explain these different national responses to the potential global governance of climate change. Although many social theories describe the relationship between society and the natural environment in the developed world, they come up short in providing a clear explanation for why some nations and not others support the international agreement to regulate greenhouse gases and slow global climate change.

In contrast to these theories, I propose the notion of the global environmental system to explain the complex interrelations of social actors involved in the tensions between domestic decisions and international environmental policy making.

At the heart of this topic is the question of how states respond to global governance and multilateralism. In the wake of the events of September 11, 2001, the post–Cold War world order has become more visible. In particular, with the fall of the Soviet Union as one of the two world superpowers, the United States has emerged as the lone economic, political, and military leader, capable of making unilateral decisions that have global implications. As Richard Falk notes in his article "Defining a Just War" (2001, 14), the Bush administration has displayed what he calls an "ingrained disdain for multilateralism." Although this disdain has become more apparent with the United States' response to the terrorist attacks on the World Trade Center and the Pentagon, it was relatively clear when Bush made his decision to reject the Kyoto Protocol unilaterally without regard to the other states that had been working on this issue for almost a decade. These events emphasize the focus of this book: Global policy making is only as effective as the states that support it.

Environmental treaties have been a relatively recent addition to the geopolitical scene. These international policies are a product of political globalization: the internationalization of regulations to defend the global environment from the environmentally damaging practices of transnational actors and nation-states through the formation of multilateral environmental regimes. Although interest in global environmental change mushroomed after the 1972 United Nations Conference on Human Development in Stockholm, Sweden, the recent push in international environmental policy making began at the United Nations Conference on the Environment and Development in Rio de Janeiro, Brazil—the Earth Summit—in 1992. Five major international environmental agreements were signed during the conference itself. Since then, nine international environmental treaties have started the negotiation process, with the goal of limiting the amount of pollutants released into the environment and protecting different aspects of the world's natural endowment. Table 1.1 lists these international environmental treaties and the date that each was signed.

The formation of these new international regimes has had many effects on society beyond the regulations themselves. For example, the "epistemic communities," connecting scientists and policymakers around the world, developed as a product of international environmental regime formation (for a full discussion, see Haas 1989, 1990). Moreover, much of the increased international collaboration among social movement organizations working on global environmental issues focuses on the formation of such regimes (for a full discussion, see Keck

Table 1.1.   International Environmental Treaties Signed since the Earth Summit

Date Signed	Treaty
May 9, 1992	United Nations Framework Convention on Climate Change[a]
June 5, 1992	Convention on Biological Diversity[b]
September 8, 1994	Lusaka Agreement on Cooperative Enforcement Operations Directed at Illegal Trade in Wild Fauna and Flora
October 14, 1994	United Nations Convention to Combat Desertification in Those Countries Experiencing Serious Drought and/or Desertification, Particularly in Africa
May 21, 1997	Convention on the Law of the Nonnavigational Uses of International Watercourses
June 25, 1998	Convention on Access to Information, Public Participation, in Decision Making and Access to Justice in Environmental Matters
September 10, 1998	Rotterdam Convention on the Prior Informed Consent Procedure for Certain Hazardous Chemicals and Pesticides in International Trade
June 17, 1999	Protocol on Water and Health to the 1992 Convention on the Protection of Use of Transboundary Watercourses and International Lakes
May 22, 2001	Stockholm Convention on Persistent Organic Pollutants

Source: http://untreaty.un.org/ENGLISH/bible/englishinternetbible/partI/chapterXXVII/chapterXXVII.asp
[a]The Kyoto Protocol to the United Nations Framework Convention on Climate Change was signed December 11, 1997.
[b]The Cartagena Protocol on Biosafety to the Convention on Biological Diversity was signed January 29, 2000.

and Sikkink 1998; Risse-Kappen 1995; Smith, Chatfield, and Pagnucco 1997). Through these expanding networks of transnational social movements, some scholars have noted that a global civil society has emerged that pressures nation-states to promote social change, in some cases invoking what Keck and Sikkink call the "boomerang effect" (1998).

It is my contention that most international environmental treaties are rooted in and affect domestic environmental policies. Indeed, after the Montreal Protocol's entry into force in 1987, 167 countries around the world implemented national policies that banned the use of ozone-depleting chemicals produced in their countries. In other words, it is through the implementation and enforcement of these domestic laws that the success or failure of the regime will ultimately be judged. The national level is important not only because it is where the policies are ratified, implemented, and enforced, but it is also where the global epistemic communities of scientists and policymakers who work on environmental issues emerge from and continue to work. Similarly, transnational social movements sprout from national environmental organizations. In other words, the national level is the base of global environmental policy making.

With the influx of international institutions and organizations that work on the transnational level, these organizations also contribute to international environmental policy making. In many cases, these groups coordinate and organize the transnational negotiations that are the foundations of the policy-making process. At the same time, however, in the words of Haas, Levy, and Parson (1992, 2), "International conferences and institutions are only as effective as governments choose to make them. International efforts to promote environmental protection have been most effective when they enhance governmental concern, provide a forum for governments to harmonize international policies, and improve national capacities to cope with environmental threats."

Environmental treaties are internationally coordinated agreements that are ratified and implemented on the national level to regulate global society's consumption of environmental goods and its production of environmental bads. They are the product of negotiations among transnational social actors, international organizations, and representatives of nation-states on the international level; and, perhaps more important, the outcomes of these negotiations are often the result of the internal debates among the social actors within nation-states.

The actual process of international treaty making is usually long and complex. First, nation-states debate the wording of the documents before they sign it. Next, in a process that could take years, countries domestically endorse or ratify the treaty. For some countries, including the United States, ratification means that prior to domestic approval, the country must have designed legislation and must be prepared to approve implementing it. Speaking on this issue of implementing legislation, a senior U.S. State Department official pointed out that "it is not a legal requirement per se. . . . It is a matter of treaty practice (interview, official C 2001). Furthermore, before an international agreement can come into legal force and become a treaty, the governments of a certain percentage of the countries that have signed the agreement must also have ratified it. In the case of the Kyoto Protocol, which regulates countries' emissions of the gases that contribute to global climate change, the legislation enters into legal force ninety days "after it has been ratified by at least 55 Parties to the Convention, including developed countries accounting for at least 55% of the total 1990 carbon dioxide emissions from this industrialized group" (Climate Change Secretariat 1998, 1). Due to the importance of national ratification and implementation to the success of environmental treaties, the outcomes of this and other international environmental agreements are, in many ways, determined within nation-states.

This level of attention to the domestic side of international environmental policy making, in conjunction with the occurrences on the international stage, differs from many other approaches to international policies in important respects. First, it takes into account the complex interaction between the national and international levels, rather than attempting to explain all political outcomes regarding international environmental policies by looking only at the process of international negotiation and regime formation (see, e.g., Benedick 1991; Vig and Axelrod 1999; Young 1994). Although understanding these international processes is important, we cannot develop complete explanations of international environmental policy making without analyzing the national level, where ratification and implementation take place. If our intention is to understand the successes and failures of global environmental regulations, we cannot make progress by starting with objects of explanation that isolate only the international aspects of the decision-making process. We must look at the policy and regime formation in their complexity, as an iterative process that takes place between the domestic and international levels.

The importance of the relationship should not be overstated; some scholars of international policy making have begun to focus on the domestic level. For example, in his work on diplomacy, Putnam (1988) discusses what he calls the "two-level game" between domestic and international actors. This game is further explored in the more recent work on what Evans, Jacobson, and Putnam call "double-edged diplomacy" (1993). Similarly, in her book on the domestic sources of international environmental policies, DeSombre (2000) focuses her attention on the processes through which certain domestic environmental policies in the United States have become internationalized. Although these scholars look at the domestic side of international environmental policy making (see also, Schreurs and Economy 1997; Kawashima 2000; Keohane and Milner 1996; Sprinz and Weiss 2001), they do not probe the depths of the interactions among domestic social actors working to affect a country's ultimate position on an issue. Even though these works explore, to some degree, both the national and international levels, the focus tends to remain on isolated international political outcomes.

Rather than look only at the number of signatories or at the duration of the treaty, as many other scholars have done (e.g., Downs, Danish, and Barsoom 2000), I propose that successful environmental transformation—the actual change in the material good—be considered part of the definition of international environmental policy making. This rationale is grounded in my belief that a successful international environmental policy will effectively regulate the consumption of environmental goods or the production of environmental bads. Because I intend to focus exactly on this question in my comparative analysis—in which different nation-states' responses to the threat of international environmental regulations determine the political and material outcomes of the international policies themselves—my concept of international environmental policy making necessarily highlights, as a basic defining feature, the interaction among domestic actors working on these issues and on the outcome of the international negotiation and policy making.

TOWARD AN UNDERSTANDING OF
INTERNATIONAL ENVIRONMENTAL POLICY MAKING

How, then, are national responses to international environmental policies and their effects on the international policy-making process to be explained? Where are we to turn for useful modes of analyzing their political and material outcomes, both nationally and internationally? The only way to answer these questions is to explore the domestic side of international environmental policy making—where the actual ratification and implementation take place. Given this book's focus on the decisions taking place inside the nation-state regarding the ratification and implementation of international environmental treaties, I now turn to the more general literature on the relationship between society and the natural environment as the starting point for answering the questions.

Although these existing social scientific theories that look at society–environment relationships and their environmental outcomes are a good place to begin the analysis that I will undertake, they are not entirely adequate. Accordingly, the chief purpose of this chapter is to introduce and defend principles and methods of analysis that represent alternatives to those shared by most existing approaches. I argue that, in contrast to the processes of explanation used by the currently prevalent theories of society–environment relationships, the outcomes of international environmental policy making should be analyzed through an integrative process that focuses on the nation-state, paying special attention to national contexts and to developments on the national level that affect the states' positions within international negotiations and the characteristics of the international environmental policy under review. Furthermore, I argue that comparative analysis, employing both quantitative and qualitative methods, is the most appropriate way to develop explanations of international policy outcomes that are nationally grounded and generalizable beyond unique cases.

To make possible the following presentation of these theoretical and methodological alternatives, it is helpful to identify major types of social scientific theories of society–environment relationships, briefly sketching the important characteristics of each. It is important to note, however, that the purpose of this book is not the academic enterprise of general theory building. Instead, this book is an in-depth and systematic analysis of a set of cases, and it is concerned not only with describing the cases separately but with understanding and explaining the generalizable reason at work in the entire set of national environmental decisions under discussion and the international implications of these decisions. By briefly summarizing the extant theories of society–environment relationships, we are provided with an efficient way of identifying relevant and vital theoretical issues for later discussion.

Therefore, I suggest that the relevant social scientific theories of society–environment relationships can be grouped into two major schools. First, however, we must note that recent sociological work has handled the potential implications of constraints to society's consumption of environmental goods and emissions of environmental bads under the conditions of advanced modernity in starkly different ways. First is the research that is often categorized as being within political sociology. Mostly European, this research includes an important commonality across three main branches of work—namely, reflexive modernization, ecological modernization, and postmaterialism. All three branches involve an expectation of the emergence of an "environmental state":[1] the view that advanced or industrialized nation-states will include environmental protection "as a basic state responsibility" (Frank, Hironaka, and Schofer 2000b, 96), with the state at least implicitly having enough autonomy or capacity to carry out such a responsibility. The second body of work is the generally U.S.-based research on "environmental sociology," which has been accumulating over the past several decades and has virtually the opposite expectations, that is, looking rather pessimistically on environmental reform. Much of the mainstream theoretical environmental sociology tends to consider environmental problems to be a relatively direct consequence, or at least a clear correlate, of industrialization and capitalist accumulation.

In large part, the differences between the "environmental state" and the "environmental sociology" lines of thought can be traced to these significantly differing expectations about the nature of the relationship between state regulation and the economy. Much of the recent work on the environmental state reflects the view that environmental protection is, in the words of Anthony Giddens, "a source of economic growth rather than its opposite" (1998, 19). This optimism about the economy in turn reflects the tendency of this newer work to concur with the differences that theorists such as Habermas (1970, 1975) have seen between "liberal" and "advanced" stages of capitalism: a change in the relationship between the economy and the state (for a full discussion of these similarities, see Fisher 2002). The newer work, in other words, tends to share Habermas's view that "the continuing tendency toward disturbance of capitalist [economic] growth can be administratively processed and transferred, by stages, through the political and into the socio-cultural system" (1975, 40). In the more concise

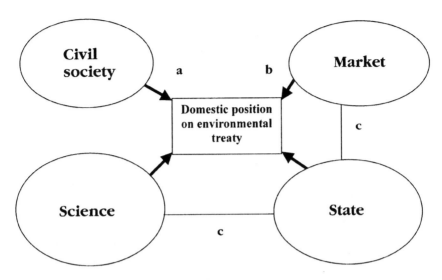

Figure 1.1. The Role of Different Social Actors in Environmental Decision Making

Expectations of Extant Literature:

a *Reflexive modernization and postmaterialism:* Civil society leads the way in determining political outcomes.

b *Environmental sociology literature:* Market leads the way in determining political outcomes.

c *Ecological modernization:* Market, science, and state work together to determine political outcomes.

assessment provided by McCarthy (1978, 363), this change means that the economy "no longer has the degree of autonomy" that it once had.

Figure 1.1 presents a simple model of the relationships among domestic actors in the formation of decisions regarding environmental policies. As the diagram illustrates, there are four major social sectors that are involved in determining a nation-state's position on an international environmental agreement: state, market, science, and civil society. Each letter *(a, b, c)* represents a social sector's effect (independent or combined) on political outcomes, as expected by the extant literature.

Theories of Environmental Sociology

Habermas's views regarding the autonomy of the economic sphere in advanced capitalist states, however, are in sharp contrast with the expectation that is evident in much of the work in environmental sociology over the past several decades. That research reflects a perspective that is more consistent with the earlier sociological work put forth by scholars such as

Edelman (1964), O'Connor (1973), and Block (1987). Specifically, much of the environmental sociology scholarship expects state constraints on business/economic "autonomy" to be relatively problematic, with the legitimacy of the state being seen as dependent, to a significant degree, on the maintenance of "business confidence" and economic growth. In addition, the earlier environmental sociology literature has tended to see environmental regulations as being, to varying degrees, antithetical to continued economic growth.

Scholars working within this perspective tend to find the market to be the leading force in society (see figure 1.1b). Important authors within the earlier body of work on environmental sociology, for example, have posited a need to stop the expansion of capitalism and the processes of industrialization to avoid the "overshooting" of global carrying capacity (see, e.g., Catton 1980); or the collapse of economic activity that would otherwise result from the self-exhausting tendencies of a "treadmill of production" (see, e.g., Schnaiberg 1980), or the "second contradiction of capitalism" (see, e.g., O'Connor 1991; Foster 1992).

Perhaps the most explicit in elucidating the connections between environmental destruction and economic prosperity are Schnaiberg (1980) and Schnaiberg and Gould (1994). Their focus on the notion of a treadmill of production posits an ever-growing process of production and accumulation, in which industries harm the environment as part of the search for growing prosperity. In the more recent work that emphasizes what they call "the enduring conflict" between the environment and the economy, Schnaiberg and Gould (1994) predict that, ultimately, environmental destruction will become the limiting factor in economic growth. In other words, the environmental destruction that they see as being a necessary component of economic growth will eventually lead to an erosion of the natural resource base such that future economic growth becomes impossible.

Similarly, O'Connor (1991) is relatively explicit in stating that natural resource consumption fuels our society's economic growth. In his words, the basic cause of the "second contradiction of capitalism" is "capitalism's economically self-destructive appropriation and use of . . . external nature or environment" (O'Connor 1991, 108). In other words, the contradiction is that the expansion of capitalism is inherently associated with increasing consumption of natural resources; but since such resources are finite, the expansion will inevitably consume all available environmental resources, which will result in environmental destruction and resource exhaustion and thus ultimately lead capitalism to collapse. O'Connor does not state that the destruction of the environment automatically makes people rich; instead, like Schnaiberg and Gould, he says that capitalist society destroys the environment to get rich.

Among other contributions, Foster clarifies and builds on O'Connor's work on the second contradiction of capitalism, describing this contradiction as the "absolute general law of environmental degradation under capitalism" (Foster 1992, 77; see also, Foster 2000). Foster is much clearer than O'Connor in depicting the depletion of natural resources as a precondition for economic prosperity. Still, it is worth noting that Foster does not actually state that environmental degradation causes economic prosperity; instead, he claims that the destruction is an inherent part of economic growth, or the "overall toxicity of production" (1992, 79). In Foster's words, the second contradiction can be "expressed as a tendency toward the amassing of wealth . . . and the accumulation of conditions of resource-depletion, pollution, species and habitat destruction, urban congestion, overpopulation and a deteriorating sociological life-environment" (1992, 78–79).

In contrast to the environmental sociology theorists who have developed arguments surrounding the implications of capitalism, Catton focused on the notion that overall levels of economic activity may have already "overshot" the planet's carrying capacity (see especially, Catton 1980). The essence of what he calls "overshoot" is that society's present prosperity is supported by higher levels of resource consumption than can be maintained indefinitely—an argument that implicitly reflects the assumption that natural resource extraction sustains economic well-being, if only in the short-to-medium run (for a review of the literature on the relationship between natural resource extraction and economic well-being, see Fisher 2001). Once the resources have run out, society will no longer be able to grow economically, and instead it may be faced with what Catton calls a "die-back."

Similarly, the work by Dunlap and Van Liere (1978, 1984), which spells out the contrasts between what they call the human exceptionalist paradigm (HEP) and the new ecological paradigm (NEP), looks at the association between economic growth and environmental degradation. In essence, the human exceptionalist paradigm involves an emphasis on the importance of economic growth, even in cases where such growth clashes with ecological conservation. Within this paradigm, as Dunlap and Van Liere describe it, humans are assumed to be exempt from the limitations of the natural environment, with their being able to continue consuming natural resources and polluting in order to sustain economic growth. By contrast, one of the basic tenets of the new ecological paradigm is that the preservation of natural resources may require the imposition of limits on economic growth. This new paradigm, which has yet to be realized, includes the notion that the earth's resources are limited and that a society that is reliant on the consumption of these resources is not sustainable.

Theories of the Environmental State

The differing degrees of optimism between the environmental sociological studies described here and the work on the environmental state can be
traced to hypotheses in the latter works, which expect advanced societies
to move toward the overcoming of their tendencies to create environmental crises—so much so that the economic prosperity of the nation-
state may well be associated with lower levels of environmental input and
emissions (for analyses and supportive evidence, see Templett and Farber
1994, Repetto 1995, and Freudenburg 1992; for opposing views, see Daly
1990 and Bunker 1996). Still, although the overall outcomes are expected
to be reasonably compatible across the three lines of work that make up
the literature regarding the environmental state—reflexive modernization, ecological modernization, and postmaterialism—the specific mechanisms that are hypothesized to support the progression toward those
outcomes, and the ways in which actors are expected to reorient themselves to deal with environmental problems, do tend to be different.

Perhaps the theorists who work within the framework of ecological
modernization are the ones who provide the most explicit explanation of
the actual processes. For example, the best-known proponents of ecological modernization in the English-language literature expect "unproblematic use of science and technology in controlling environmental problems"
(Mol and Spaargaren 1993, 454).[2] The work on postmaterialism, mean-
while, tends to emphasize the importance of changing citizen views;
perhaps the most well-known proponent of postmaterialism finds that
nation-states with "relatively post-materialistic publics" are the ones that
"rank relatively high in their readiness to make financial sacrifices for the
sake of environmental protection" (Inglehart 1995, 57).[3] Finally, proponents of the reflexive modernization perspective see civil society as becoming a driving force for environmental policy making in an age of risk.
In the words of Beck, "the goal is not a turning back but rather a *new modernity*, which would demand and achieve self-determination, and prevent its
truncation in industrial society" (Beck 1995, 17, emphasis in original).[4] In
other words, although all three theories come to the conclusion that advanced nation-states will succeed in protecting the environment, they see
different social sectors as leading the way.

In addition, theories of the environmental state identify the formation
of new coalitions in advanced capitalistic societies. Mol and Spaargaren,
for example, suggest that ecological modernization involves changes in
the traditional roles of actors, specifically including "transformations in
the role of the nation-state: . . . [with] more opportunities for non-state actors to assume traditional administrative, regulatory, managerial, corporate, and mediating functions of the nation-states" (Mol and Sonnenfeld

2000, 6–7). This expectation is mirrored in Beck's work on "subpolitics" (1997). Continuing to discuss the characteristics of reflexive moderniza- tion and the risk society, Beck claims that, with the emergence of new types of conflict, "new coalitions become thinkable" (1997, 52).

Although these theories of the postindustrial society–environment re- lationship have similarities, they come to strikingly dissimilar conclu- sions regarding the roles of different societal actors in bringing about a more environmentally sound future. Some of these differences, in fact, have been outlined by Mol and Spaargaren themselves (1993). Ecologi- cal modernization theory includes the notion that the quality of the environment can and will be improved through, in the words of Buttel, "'superindustrialization'—albeit industrial development of a far differ- ent sort than that which prevailed during most of the twentieth century" (Buttel 2001, 18). Ecological modernization theorists discuss the role of the state, science, civil society, and economic spheres in postindustrial society, but they tend to focus on the state, science, and economic (or market) actors as the leaders of environmental change (see figure 1.1c). In other words, although ecological modernization theory points to an increased role for "non-state actors" (Mol and Spaargaren 2000, 3), much of the attention is paid to economic actors, such as private companies. In the words of Spaargaren and Mol (1992, 341), ecological modernization focuses on the "industrial dimensions of modernity." The theory does not overlook the role of civil society nor the role of social movement organizations, such as nongovernmental organizations (NGOs; see Mol 2000a), but it generally looks to "modern institutions such as science and technology and state intervention" to lead the way (Mol and Spaargaren 1993, 454–55).

In contrast to ecological modernization's top-down approach to social change, both reflexive modernization and postmaterialism have a more bottom-up orientation. Turning first to reflexive modernization, environ- mental improvements are seen as being the product of social responses to the increasingly egalitarian distribution of environmental risks. Most scholars working on reflexive modernization focus on civil society as the actor that drives environmental change. In Beck's words, it is through the distribution of risk that "new opportunities for arranging society arise un- der that pressure of the industrial threat that humanity will annihilate it- self and the breakup of social classes and social contrast that it causes" (1995, 2). Beck describes the implications of these social changes in his more recent work, calling the effects of the universalization of risk de- scribed here "industrial democracy" (1997, 53). Reflexive modernization theorists discuss states and economies as actors, but they indicate that the state and economic actors "need to be supplanted by an increasing par- ticipation on the part of citizens whose actions and self-binding are oriented toward enlightenment, solidarity, and responsibility" (Offe 1996, 34).

Like reflexive modernization, postmaterialism tends to be seen as involving social change that comes from civil society (see figure 1.1a). In contrast to ecological modernization and reflexive modernization theories, however, the work on postmaterialism tends to focus on individual behaviors within advanced capitalist nations. Abramson and Inglehart summarize the postmaterialism thesis: It "assumes that the economic security created by advanced industrial societies gradually changes the goal orientations of mass publics" (1995, 9). Although the postmaterialism thesis has come under criticism by scholars looking at global environmentalism,[5] many researchers use the notion of postmaterialism to study the attitudes and role of civil society in industrialized countries.[6] In general, scholars find that people with postmaterialist values are more apt to prioritize environmental protection. Postmaterialism is similar to reflexive modernization in its concentration on the role of civil society; however, rather than focus on social movement organizations and civic associations, the research on postmaterialism looks at lifestyle issues and consumer behavior as mitigating environmental protection.

AN INTERRELATIONAL PERSPECTIVE

Even with such brief sketches of the theories of society–environment relationships, it should be readily apparent that there are enormous disagreements between the environmental sociology and environmental state theories, not only about how to explain environmental policy outcomes, but also about the overall success of capitalist society. Stepping back from these larger disagreements, however, I find it important to note the lack of consistency among these theories' foci. In other words, there is not any consensus regarding which social sector is responsible for successful or unsuccessful environmental policy outcomes. In addition, theories within these schools of thought have very little to say about the interrelation between the domestic and international levels, which has recently become the norm in dealing with global environmental issues through international environmental treaties. Thus, existing theories of society–environment relationships unfortunately prove inadequate for explaining the complex interrelations of social actors involved in the tensions between domestic decisions and international environmental policy making. None of the theories adequately explain the complexity of countries' domestic positions on the regulation of global environmental problems, nor do they explain why national responses to international environmental policy making may differ. In this book, I integrate these multiple perspectives of the relationships between society and the natural environment to understand nation-states' responses to the threat of the global governance of the environment.

Even though much of the past research has addressed one or more social sectors, none explicitly examines the interrelation of these forces. To understand the relationship between domestic actors and international environmental political outcomes, a broader orientation is required. The concept of the global environmental system, described in detail in the pages that follow, provides a framework for the complex interrelations involved in the current challenges to the global governance of the environment posed by the domestic interests of nation-states. This book provides both empirical elaboration and theoretical development of this framework, specifying the ways in which the relationships between national and global actors are set within the context of the continual interactions of scientific, governmental, corporate, and citizen-led forces.

Domestic Actors within the Global Environmental System

Depending on the specific configurations of social sectors in each nation-state, different actors drive the decision-making process regarding the state's national position in international environmental negotiations as well as the domestic responses to an environmental treaty. To restate, an adequate understanding of the relationship between domestic actors and international environmental political outcomes demands a full consideration of the roles of science, industry, civil society, and the state; and the interaction among them. Figure 1.2 portrays a model of the interactions among domestic actors in the formation of national positions regarding international environmental policies. As can be seen by the overlapping sections, this figure highlights the interrelations among relevant actors and points to the fact that the configurations of each domestic regime is important. In other words, domestic actors involved in the formation of each country's respective environmental regimes arguably work in collaboration and, on some occasions, in conflict with one another: it is the interaction among the social actors in each nation-state that explains the variable political responses to international environmental policy making.

Overview of the Global Environmental System

Although it is necessary to look at the interrelations among domestic social actors to understand a country's position on international environmental policies, it is also important to conceptualize these relationships within a broader global environmental system; otherwise, important aspects of international environmental regimes will be missed. To understand fully the political outcomes of each nation-state, it is necessary to

Chapter 1

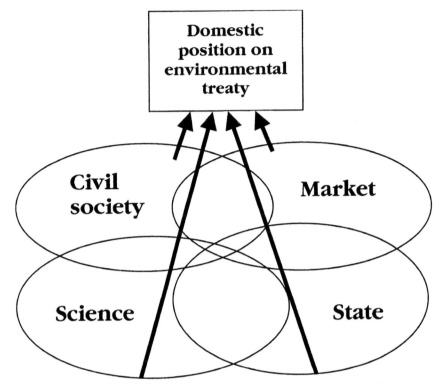

Figure 1.2.   Domestic Actors within the Global Environmental System

recognize that, although each nation can be viewed as a separate and comparable case, these countries are all part of an interactive global system. Perhaps McMichael best summarizes the relationships among cases when he says that "outcomes . . . may *appear* individually as self-evident units of analysis, but in reality are interconnected processes" (emphasis in original, 1990, 396). In other words, each country's position on international environmental treaties is decided not just through the mediation of interrelated domestic actors but also through each country's interactions with other parties and international organizations working on the environmental regime. In addition, it is important to note that these decisions are embedded in each nation's history and position within the global system in terms of economic, political, and environmental characteristics.

Figure 1.3 provides a depiction of the global environmental system. The figure illustrates the relationships among the international and domestic actors involved in the decisions regarding international environmental policies. The interactions among global actors, domestic actors, and

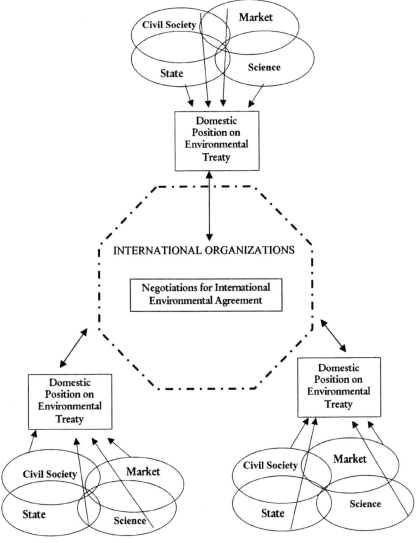

Figure 1.3. The Global Environmental System

international organizations, implicit to the global environmental system, provide more complexity to our understanding of international decision making about environmental and other social issues within an increasingly globalized world. Thus, by examining the formation of an environmental treaty through the lens of the global environmental system, I provide an account of how domestic politics have serious global implications. In other words, what happens inside nation-states in response to international political issues is becoming increasingly more important, rather than less so.

TESTING THE FORMATION OF AN ENVIRONMENTAL TREATY

Presently, the expectations deriving from the research that I classify as environmental state literature already have a degree of empirical support, particularly in studies of international institutions[7] and within the context of an emergent world polity.[8] At the same time, however, the work on international institutions and the environment has come under criticism. Of particular relevance in the current context is the criticism put forth by Buttel (2000a, 118), namely, that a focus on the establishment or the diffusion of institutional forms of environmental protection may actually have little to say about the extent to which such measures or forms "have, or are likely to have, any definite connections with actual environmental protection outcomes" (Buttel 2000a, 118).

How, then, can a sociologist hope to develop valid explanations of these actual environmental protection outcomes? It is my contention that the only way to analyze the material and political outcomes of international environmental policy making is through a comparative approach that employs quantitative and qualitative analyses. Accordingly, this book proceeds by looking first at an overview of the relationship between characteristics of advanced nation-states and their contribution to global environmental change. This analysis provides a backdrop onto which the research regarding political outcomes can be projected. Next, the book presents a qualitative analysis of the social actors involved in the domestic environmental policy-making process in three particularly telling nation-states. In the end, this comparative approach contributes to our understanding of political and material environmental outcomes in the global environmental system.

Of all of the "actual environmental protection outcomes" that might be considered, perhaps the outcome that has received the greatest amount of attention in international circles in recent years has involved contributions to global warming through the emission of what have now come to be called "greenhouse gases"—carbon dioxide (CO_2), methane (CH_4), nitrous oxide (N_2O), hydrofluorocarbons (HFCs), perfluorocarbons (PFCs), and sulphur hexafluoride (SF_6). As discussed in the chapters that follow, this issue is both political and material: scientists have been researching and debating the causes and effects of global climate change, while politicians have been discussing possible mitigation strategies.

The major global climate change mitigation strategy under international negotiation is an international environmental agreement that was drafted to regulate the gases that have been found to contribute to global warming: the Kyoto Protocol. This often-controversial agreement went through the process of high-level international negotiation and revision from 1997 to 2002. At the same time, the measures highlighted by the protocol are, and may continue to be, resisted quite intensely by key economic actors within many individual nation-states.

Given this opportunity to follow a major scientific and political issue as it develops, this book looks at the emergent global climate change regime to address questions regarding why some states support international environmental regulation and others do not. In other words, using the case of the Kyoto Protocol, I look at what characteristics of advanced capitalist states best explain their levels of environmental degradation and their support of international regimes to protect the environment.

This book moves forward after providing background information regarding the issue of global climate change, to present empirical support for the global environmental system using the emerging global climate change regime as its case. It follows industrial countries' responses to the issue of global climate change during the four critical years following the drafting of the Kyoto Protocol, in 1997. As a first step in this study, chapter 2 summarizes the history of both the science and the politics of the issue, a history dating to the late 1970s, when it was recognized as a global environmental problem. Chapter 3 takes a broad look at the relationship between contribution to global climate change and a number of economic and sociocultural variables in advanced nation-states. Although the results suggest that a number of "environmental state" variables help to explain the variation in carbon dioxide emissions within postindustrial nations, they do so in ways that in turn suggest considerable variation within the industrialized states. Thus, chapter 3 closes with an explanation of how best to proceed, with an in-depth study of this issue through the analysis of three more intensive qualitative cases studies. Chapters 4 through 6 present the results of the case studies, which focus on the domestic climate change regimes in three particularly telling nation-states—Japan, the Netherlands, and the United States—paying particular attention to the roles that social actors from state, market, science, and civil society have been playing in each nation-state. Finally, chapter 7 compares the results of these cases by examining the overall outcome of the negotiations and the future potential of the Kyoto Protocol as a treaty to mitigate global climate change. This chapter concludes by further explaining the differences in the relationships among social actors on the national and international levels more clearly within the "global environmental system." In addition, the chapter discusses the political and environmental implications of the results of this research.

NOTES

An earlier version of this chapter was published as "Global and Domestic Actors within the Global Climate Change Regime: Toward a Theory of the Global Environmental System," *International Journal of Sociology and Social Policy* 23, no. 10 (2003): 5–30.

1. See, for example, Buttel (2000a); Frank, Hironaka, and Schofer (2000a, 2000b). See also, Goldman (2001).

2. See also Buttel (2000b, 2000c); Christoff (1996); Cohen (2000); Fisher and Freudenburg (2001); Hajer (1995); Huber (1985); Leroy and van Tatenhove (2000); Mol (1995, 1997, 1999, 2000a, 2000b); Mol and Sonnenfeld (2000); Mol and Spaargaren (2000); Spaargaren (1997, 2000); Spaargaren and Mol (1992); Spaargaren and van Vliet (2000).

3. See also, Abramson and Inglehart (1995); Abramson (1997); Brechin and Kempton (1994, 1997); Dunlap and Mertig (1997); Inglehart (1990); Kidd and Lee (1997a, 1997b); Pierce (1997).

4. See also, Beck (1987, 1997); Beck, Giddens, and Lash (1994).

5. See, for example, Brechin (1999); Brechin and Kempton (1994); Dunlap and Mertig (1997).

6. See, for example, Abramson (1997); Inglehart (1995); Kidd and Lee (1997a, 1997b); Pierce (1997).

7. See Frank, Hironaka, and Schofer (2000a, 2000b). See also, Haas (1989, 1990, 1995); Haas and Sundgren (1993); Levy, Keohane, and Haas (1993); Young (1989, 1997).

8. See, for example, Frank (1997, 1999); Meyer (1994); Meyer and others (1997).

2

The History of the Science and Policy of the Global Climate Change Regime

The purpose of this chapter is to provide a short summary of the research and international policy making to date on the issue of global climate change. I begin by providing a brief overview of the issue, and I follow by discussing the relevant scholarly research. I briefly describe the themes within the natural science literature on climate change; however, given this book's focus on the social side of the issue of climate change, this chapter allots more space to the social scientific research about climate change. Finally, to provide a framework for understanding this research, I describe the political history of the issue within the global arena, paying particular attention to the status of the environmental regulation tool presently under negotiation: the Kyoto Protocol.

Since the 1970s, the biophysical phenomenon called *global warming*, or global climate change, has become one of the most prominent global environmental issues, receiving growing scientific and political attention in the past thirty years. Within the scientific world in particular, recent research has focused not only on the natural scientific topics related to climate change but also the issue's social aspects. Although the academic literature shares the common focus of global climate change, the characteristics studied are quite different across the relevant academic disciplines.

At the same time that the academic research on climate change has developed, states and international organizations have formed programs to address the issue. Both within nations and across, climate change study groups have produced reports on the political implications of regulating this issue. Some of the more prominent organizations working in this field include the United Nations,[1] the Organisation for Economic Co-operation and Development (OECD),[2] the International Energy Agency (IEA),[3] and the Global Environment Facility.[4]

With this increased attention, the topic of climate change has motivated debates within the scientific world as well as the political. A *New York Times* article, in fact, called global warming "a classic example of the persistent mismatch between the language of science and the needs of policy" (Revkin 2000). The subject of climate change, like other global environmental issues, combines policy making and science, but it does so in specific ways. "The problem of global warming elicits less scientific and political consensus than does the problem of ozone depletion. In addition, the technical options available to address the global-warming issue are less clearly defined and more diffuse, while the economic and social costs promise to be higher" (Miller and McFarland 1996, 55). The economics of mitigating global climate change has been debated quite sharply in policy circles at the national, regional, and global level; as spelled out in my case study in chapter 6, the debates in the United States have been especially heated.

In addition, the scale of the global climate change issue has at least three significant characteristics that make it unique. First, climate change is an issue that is truly global, as the atmosphere does not respect sociopolitical or economic boundaries. In the words of Rosa and Dietz, for example, the issues associated with global warming represent the "pure case of a collective good" (1998, 437); alternatively, in the nearly poetic language of Soroos, the issue of global climate change has left "humanity in the throes of a 'tragedy of the atmosphere'" (2001, 2; see also Soroos 1997, 1998). Second, although the scale of the problem is global, the effects of climate change will be local (and, in some cases, they already are); furthermore, these effects will not be consistent across political and geographical boundaries. In the state of Wisconsin, for example, ice fishers have been dissatisfied with the recently shortened fishing season; and cross-country skiers involved in the "largest, most prestigious cross country ski marathon in North America," the American Birkebeiner, were dismayed when the "Mother Nature's Birkie 2000" was canceled due to lack of snow.[5] Yet the third issue of scale is perhaps the most significant: the time scale of the issue is such that the full effects and extent of climate change will not be felt for generations (for a full discussion of the intergenerational aspects of the issue, see Thompson and Rayner 1998). In fact, even if the Kyoto Protocol were to be promptly implemented, experts such as Hugh Pitcher, staff scientist of the global change group of the Pacific

Northwest National Laboratory, claim that with our best understanding of the science today, "we are going to see somewhere between 40 and 60 cm of sea level rise by 2100"—enough of a rise to make a number of small island states uninhabitable. In Pitcher's view, "You know, it's callous, it's hard-hearted, but [the issue is one of] a few hundred thousand people against turning a global system on its head in a very short period of time. And that's really what we would have to do to protect the shoreline. [It is an] unfortunate aspect of reality" (interview, April 27, 2000). As this scientist from a national lab in the United States points out, global policy making to mitigate climate change is a difficult issue.

In other words, one of the major problems with the issue of global climate change is that "by the time the impact becomes too clear to debate, it will be far too late to do anything about it" (Revkin 2000). This fact has led some political actors to push for what is called the *precautionary principle.* O'Riordan and Cameron (1996) describe the principle:

> International agreements covering global or regional environmental protection are increasingly based on proactive or preventative measures which encompass collective action and burden sharing. As a result signatories to major agreements have to give [in] to significant changes in their economies and lifestyles ahead of scientific proof of the likely gains of making such sacrifices. This is one of the most awkward aspects of applying the precautionary principle in a democracy. (15)

Interpretations of the precautionary principle, however, are by no means universal. As spelled out in the chapters that follow, there continues to be significant disagreement about how nation-states should adopt the precautionary principle with respect to the issue of global climate change.

Even with scientists from the United States and abroad predicting significant environmental and social changes as a result of global climate change, the policy world has been slow to regulate the gases that have been identified as contributing to the greenhouse effect. Instead, there have been ongoing debates surrounding the science and policy of the issue, often focused on the question of how to regulate this global phenomenon that will surely affect nation-states in significantly different ways. In the meantime, the negotiation of a viable and enforceable international environmental treaty has been progressing only slowly, if at all.

THE SCIENCE OF GLOBAL WARMING

Since global climate change became a political issue on the international stage, a number of reports have been published that challenge the science of climate change (e.g., Mendelsohn 1999; Michaels 1992; Singer 1998). In the words of Bjørn Lomborg's *Skeptical Environmentalist*, which received significant attention after its publication in English,

> We need to separate hyperbole from realities in order to choose our future optimally. . . . Global warming will not decrease food production, it will probably not increase storminess or the frequency of hurricanes, it will not increase the impact of malaria or indeed cause more deaths. It is even unlikely that it will cause more flood victims. (2001, 317; but see Grubb 2001; Pimm and Harvey 2001; Schneider 2002)

Beyond Lomborg's work, others have focused on the notion that global warming is not necessarily the product of human influences on the world's climate (e.g., George C. Marshall Institute 2001; New Hope Environmental Services 2000). In addition, some studies have even found that increased carbon dioxide "will confer a net benefit on society" (e.g., New Hope Environmental Services n.d., 1; Idso 1997). Although these reports have received a lot of attention within the United States and have been covered by major American media outlets, the overwhelming majority of peer-reviewed scientists working around the world who study global climate change, including most of the scientists in the United States (National Research Council 2001), have reached a general level of consensus about the issue.

In fact, the Intergovernmental Panel on Climate Change (IPCC), which was established to synthesize and summarize the research on global climate change, has come to starkly different conclusions. Robert Watson, the chair of the IPCC from 1996 to 2002, summarized the state of the science at the November 2000 round of the climate change negotiations: "The overwhelming majority of scientific experts, whilst recognizing that scientific uncertainties exist, nonetheless believe that human-induced climate change is inevitable. . . . The question is not whether climate will change in response to human activities, but rather how much (magnitude), how fast (the rate of change) and where (regional patterns)" (Watson 2000, 2). Watson echoes the statements put forth by the IPCC's third assessment report, which finds that "additional data from new studies of current and palaeoclimates, improved analysis of data sets, more rigorous evaluation of their quality, and comparisons among data from different sources have led to greater understanding of climate change. The global average surface temperature has increased over the 20th century by about 0.6°C" (IPCC WGI 2003, 2).

To understand this growing scientific consensus, it is helpful to clarify what is meant by "global climate change" in the technical literature. For most research on the subject, the phenomenon of global climate change is generally understood as involving a warming scenario. Although the majority of the current predictions come in terms of an increase in average temperatures, the magnitude of the increase will vary across the globe, and researchers point out that some areas may even experience temperature decreases. Further, some scientists predict that a change in climate variability, and hence an increase in extreme events such as storms, may also play an important role in climate change, thereby resulting in additional ecological and social responses (see, for example, Barrow and Hulme 1996; Joubert and Hewitson 1997; Michaels et al. 1998). The research on this subject, however, has yielded conflicting results, with changes in variability proving difficult to predict. As such, "the scientific debate on this dimension of global change [variability] continues" (Michaels et al. 1998). To date, most of the natural and social science literature thus far speaks about climate change in the form of global warming.

Conclusions of the Intergovernmental Panel on Climate Change

The issue of global climate change, or "global warming," perhaps first became a truly major research topic with the First World Climate Conference in 1979. That scientific conference focused on how climate change may affect human activities. During the 1980s and 1990s, a number of international conferences took place that included the participation of scientists, governments, and policymakers. In 1988, and partly in response to differing scientific views, the IPCC was formed by the World Meteorological Organization (WMO) and the United Nations Environment Programme (UNEP) to "assess the state of existing knowledge about the climate system and climate change; the environmental, economic, and social impacts of climate change; and the possible response strategies."[6] To restate, the purpose of the IPCC is not to conduct original scientific research; rather, the IPCC was set up to synthesize and summarize the research taking place around the world. Beyond scientific assessments, the IPCC was designed "to formulate realistic response strategies" to global climate change (Paterson 1996, 43). In addition to producing assessment reports on the state of the science of climate change, the IPCC process also involves the release of a "summary for policymakers," created by the authors of each assessment review and approved by nation-states involved in the process. These summaries are considered key to the process. In the words of Robert Watson, the "power of the IPCC is the intergovernmental process through the summaries for policymakers" (interview, July 23, 2001). The IPCC now involves the work of over 2,000 of the leading climate scientists in the world, divided into three working groups (WGs): science (WGI), impacts (WGII), and responses (WGIII). In 1998, the IPCC added the Task Force on National Greenhouse Gas Inventories to "develop and refine an internationally agreed methodology and software for the calculation and reporting of national GHG emissions and removals."[7]

Given the mission and stature of the IPCC, any discussion of the research on climate change is best served by a summary of the IPCC's findings in their recent review. The IPCC periodically produces an assessment of the literature on climate change, each of which is peer-reviewed. The first such assessment review was released in 1991; the second in 1996; and the third in 2001, which was divided into three subreports, each authored by a different working group (respectively, IPCC 1991, 1996; IPCC WGI 2001; IPCC WGII 2001; IPCC WGIII 2001). The following section provides a brief summary of the IPCC's current findings as outlined by the chair of the IPCC, Robert Watson, at the climate change negotiations in November 2000 (Watson 2000).

Working Group I

The IPCC Working Group I focuses its review on the science of climate change; it does so by looking at the earth's climate system and the influence of human activities. The major conclusions of the recent assessment by this working group are as follows:

1. The earth's climate is changing.
2. The atmospheric concentrations of greenhouse gases are changing due to human activities.
3. The weight of scientific evidence suggests that the observed changes in the earth's climate are due, at least in part, to human activities.
4. The emissions of greenhouse gases are projected to increase in the future, due to human activities.
5. The latest projections of carbon dioxide emissions are consistent with earlier projections, although projected sulfur dioxide emissions are much lower.
6. The global mean surface temperatures are projected to increase by approximately 1.5 to 6.0 degrees Celsius by 2100.
7. The seasonal and latitudinal shifts in precipitation show that arid and semiarid areas are becoming drier; sea levels are projected to rise about fifteen to ninety-five centimeters by 2100.
8. The frequency and magnitude of the "El Niño" events (generally described by scientists as El Niño-Southern Oscillation [ENSO] events) may increase.
9. The incidence of some extreme events are projected to increase (for a full discussion of these findings, see IPCC WG1 2001).

Working Group II

The second working group of the IPCC focuses its attention on the scientific research on the impacts of global climate change. This working group concentrates on the vulnerability of water resources, agriculture, natural ecosystems, and human health to climate change and sea-level rise, for different regions of the world and at regional and global scales. This working group's recent assessment is as follows:

1. Climate change could exacerbate water stress in arid and semiarid areas, and most regions will experience an increase in floods.
2. Biological ecosystems have already been affected by changes in climate during the last several decades.
3. Agricultural productivity is projected to decrease in many countries, especially in the tropics and subtropics.
4. Climate change is also projected to alter the structure and functioning of ecological systems and to decrease biological diversity.
5. Forests and coral reefs are especially vulnerable to increases in temperature.
6. Human health is sensitive to changes in climate because such a scenario implies changes in food security, water supply, and quality, as well as the functioning and range of ecological systems.

(For a full discussion of these findings, see IPCC WGII 2001; for more specific research on the natural scientific aspects of climate change see, e.g., Adams et al. 1998; Bradley et al. 1999; Carpenter et al. 1992 ; Davis and Shaw 2001; DeGroot, Ketner, and Ovaa 1995; Hogg and Williams 1996; Jones, Osburn, and Briffa 2001; Lambeck and Chappell 2001; Magnuson et al. 1997; Schindler 1997; Schwartz 1994; Sparks and Carey 1995; Winnet 1998; Zachos et al. 2001.)

Working Group III

In contrast to the majority of the themes within Working Groups I and II, Working Group III tends to look at social issues. In particular, Working Group III concentrates on research into the mitigation of climate change through reduced emissions and enhanced carbon sinks, focusing on the ability of the terrestrial biota—and to a lesser extent, the oceans—to trap carbon. The majority of the research reviewed by this working group, however, is social scientific in nature. The main conclusions of this working group are as follows:

1. Significant reductions in net greenhouse gas emissions are technically and economically feasible.
2. Policy instruments can be used to facilitate the penetration of lower carbon-intensive technologies and modified consumption patterns.
3. Energy services are critical to poverty alleviation and economic development.

4. Secondary benefits, such as development and equity, can lower the cost of climate change mitigation.
5. Significant reductions in greenhouse gases can be accomplished by the global community's pursuing sustainable development goals.
6. Technology transfer is a critical issue to the mitigation of global climate change (for a full discussion of these findings, see IPCC WGIII 2001).

Social Science Literature on Global Climate Change

In addition to the brief overview provided by the IPCC, the effects of climate change on social systems and the roles of social structure in climate change mitigation strategies have received some attention (see, e.g., the assessment by Crenshaw and Jenkins 1996; see also, Rayner and Malone 1998). Outside the field of economics, the social scientific literature on this subject tends to follow five major themes: the politics of climate change, the science of the issue, the relationship between emissions and the social characteristics of nation-states, social perceptions, and the media coverage of global warming. Given the relevance of the social science research to the present project, each of these themes is addressed in turn.

Politics of Climate Change

With international negotiations of the Kyoto Protocol taking place annually, it is not a surprise that a significant portion of the social science literature on climate change tends to focus on policy issues. This literature tends to fall into two main categories. The first involves work that is done from an international relations perspective, building off the literature on international organizations and the global governance of environmental problems.[8] It includes academic research that discusses the potential for, and outcomes of, international regulation to mitigate global warming.[9] Research in this vein also includes analysis of the international negotiations themselves (see, e.g., Gelbspan 1997; Schroeder 2001), and much of it looks at the mechanisms involved in the Kyoto Protocol. Messner and colleagues (1992, 6), for example, focus their discussion on what they call the "five categories of proposals aimed at reducing CO_2 emissions." The proposals described in this piece are intended to contribute to the policy decisions about the best regulatory methods to mitigate global warming (see also, Rudel 2001).

The second type of research on the politics of global climate change focuses instead on the various actors involved in potential regulation of greenhouse gases.[10] This literature tends to look at the establishment of national positions and the international institutions involved in the negotiations. Boehmer-Christiansen (1997), for example, discusses the relationship between the World Bank, the Global Environmental Facility, and the United Nations in regulating climate change.

Outside of academia, a number of researchers have written on the politics of climate change within the more popular press. The majority of this work tries to explain the problems with the global climate change regime as it develops through international meetings.[11] The verdict is not necessarily optimistic. In the words of Sarewitz and Pielke, for example, "The enormous scientific, political, and financial resources now aimed at the problem of global warming create the perfect conditions for international and domestic political gridlock, but they can have little effect on the root causes of global environmental degradation, or on the human suffering that so often accompanies it" (2000, 2).

Science of Climate Change

The social scientific literature on the science of climate change is much less extensive. Much of this literature follows the themes set out by the U.S. National Academy of Science's work on the Human Dimensions of Global Change (see, e.g., Committee on the Human Dimensions of Climate Change 1992, 1999; Rayner 1992). Some of this research looks at the construction of the science surrounding the debate (e.g., McCright and Dunlap 2000). One such case in point is provided by the work of Rosa and Dietz (1998), who compare the neorealist and interpretive notions of the science of global climate change. Boehmer-Christiansen (1994, 142), in contrast, focuses on the question "How and with what effect did the scientific community influence climate policy?" Other scholars look at the role that is being played by the Intergovernmental Panel on Climate Change (IPCC). Wynne (1994), for example, focuses his attention on global environmental issues, using the IPCC as an example of how "'scientific discourses'. . . prevent us from recognizing and responding to the awful challenges of the larger version of the global crisis" (1994, 172; see also Jasanoff and Wynne 1998). The U.S. National Academy of Science report on the Human Dimensions on Global Change takes an altogether different position, saying that "contributions to the recent Intergovernmental Panel on Climate Change (IPCC) reports are a good illustration of the significance and policy relevance of human dimensions research" (Committee on the Human Dimensions of Global Change 1999, 6; see also Skodvin 2000).

The Relationship between Emissions and the Social Characteristics of Nation-States

A third body of social science work compares the behaviors of nation-states. In particular, it focuses on the relationship between emissions and affluence (and other social characteristics), analyzing how they are related to national emissions (Dietz and Rosa 1997; Roberts and Grimes 1997; Roberts 2001; see also Fisher and Freudenburg 2002; York, Rosa, and Dietz 2003). Roberts and Grimes, for example, use emissions to test the hypothesis that there will be an environmental "Kuznets curve"[12] and that the relationship is not the result of "groups of countries passing through stages of development, but of efficiency improvements in a small number of wealthy countries," combined with worsening performance in poor and middle-income countries (1997, 191). Dietz and Rosa (1997) come to similar conclusions, finding that the developed and developing countries follow different emission trajectories.

Perceptions of Climate Change

The social science literature on the perceptions of global climate change includes research on both the local and international levels.[13] Perhaps Dunlap (1998) provides the most extensive international study of public views of global warming. In his research comparing citizen perceptions of global climate change in six nations, he finds that "lay opinion does not differ all that much from the emerging 'consensus' view of the scientific community involved in global change research" (1998, 492). Read, Bostrom, and Granger-Morgan (1994) and Bostrom and colleagues (1994), in comparison, look at laypeople's general level of understanding of the issue at the local level. After surveying small samples of laypeople at or near Carnegie Mellon University in Pennsylvania, the authors come to different conclusions than Dunlap's larger, international study, arguing that "subjects had a poor appreciation of the facts" (Read, Bostrom, and Granger-Morgan 1994, 971). An altogether different type of study on the perceptions of the issue of climate change was published by Barrett, Chambers, and Schroeder that studies political elites (2001). This study includes data concerning UN delegates' perceptions of science and politics that were gathered at the climate change negotiations following the Framework Convention on Climate Change.

Media Coverage of Climate Change

As in the case of the research on the perceptions of climate change, the literature about the media coverage of global climate change investigates the relationship between policy processes and media coverage as a potential means of further understanding political outcomes and public opinion of the topic. Some of the research looks at the coverage in comparison to that of other environmental issues in the news (Mazur 1998; Mazur and Lee 1993; Ungar 1998). Other research focuses exclusively on the media coverage of the topic of global climate change, paying particular attention to specific issues that are reported by the media (e.g., Shanahan and Trumbo 1998; Trumbo 1996; see also Weingart, Engels, and Pansegrau 2000). Trends in media attention to such issues are often interpreted with reference to what Downs (1972) calls the "issue attention cycle," or the expectation that attention to social problems are less the result of objective conditions than of a cycle of political and media attention. One of the more interesting conclusions of this literature is reported by Shanahan and Trumbo (1998, 16), who find that, at least in the United States, the "global warming coverage cycle [is] an issue that became less certain over time, reflecting a narrative move from established science to dynamic controversy." In other words, media coverage of the issue of climate change had shifted to cover the more dramatic, politically peripheral disagreements about the science of climate change, away from the drier coverage of the established academic science. The emergence of a debate surrounding the science of climate change is particularly relevant in the case of the United States, which is discussed in detail in chapter 6.

THE HISTORY OF INTERNATIONAL CLIMATE CHANGE POLICY

The social scientific literature on global climate change is remarkably disparate in its subject matter and the units of analysis that are studied. A large part of this research, however, tends to look at the international level, following the UN policy-making process as it develops. In the section that follows, I present the significant international events surrounding the 1997 drafting of the Kyoto Protocol, as they are significant to understanding domestic responses to the international regulation of global climate change. Since published studies exist that focus on the emergence of the political issue of climate change up until the drafting of the Kyoto Protocol (see, for example, Gelbspan 1997; Leggett 1999; Paterson 1996), this summary is concise, merely providing a framework for subsequent chapters about domestic and international climate change regime formation.

Constructing Climate Change: The Early Years

The issue of global climate change entered the global arena with the First World Climate Conference, in 1979. Some analysts, however, trace the roots of the interest in this topic to the 1972 United Nations Conference on Human Development, in Stockholm (e.g., Skodvin 2000). Five years after the Stockholm meeting, the World Climate Conference was held in Geneva, during which the World Climate Program was started. This scientific conference focused on how climate change might affect human activities. Since that time, research and policy making have developed

Table 2.1.   Climate Change Policy Timeline

Year	Development
1972	UN Conference on Human Development held in Stockholm, Sweden
1979	First World Climate Conference
1988	Establishment of Intergovernmental Panel on Climate Change (IPCC)
1990	IPCC First Assessment Report Second World Climate Conference UN General Assembly approves start of treaty negotiations.
1992	IPCC is restructured, and Working Group III is redesigned to review information on socioeconomic issues related to climate change mitigation. UN Framework Convention on Climate Change (UNFCCC) signed at Earth Summit. (June)
1994	IPCC Interim Report UNFCCC enters into legal force. (March 21)
1995	Conference of Parties 1 (COP-1) held in Berlin, Germany. (March 28–April 7) IPCC Second Assessment Report approved. (December)
1996	COP-2 held in Geneva, Switzerland. (June 8–19)
1997	COP-3 held in Kyoto, Japan. The Kyoto Protocol was adopted at this meeting. (December 1–11)
1998	COP-4 held in Buenos Aires, Argentina. (November 2–13)
1999	COP-5 held in Bonn, Germany. (October 22–November 5)
2000	COP-6 held in the Hague, Netherlands. (November 13–24)
2001	IPCC Third Assessment Report approved. (January) U.S. president George W. Bush pulls out of climate change negotiations. (March) COP-6 is held in Bonn, Germany. (July 16–27)

around the subject of global climate change. Table 2.1 presents a summary of the major scientific and political events in the history of the development of a global climate change regime.[14] Especially for chapters 4 through 6, it may be useful to refer back to this timeline.

As discussed, the IPCC was formed in 1988. Scholars such as Paterson (1996, 47–48) report that when the IPCC's first assessment report was released in 1990, it "set the stage for the Second World Climate Conference. . . . The scientists involved sent a clear message to politicians that they believed the latter should act to reduce the threat of global warming." Later, in 1990, the Second World Climate Conference was held and attended by officials from several international organizations, including the United Nations, as well as representatives from 137 nation-states. The conference included negotiations among the states in attendance and called for a framework treaty on climate change. The United Nations General Assembly approved the start of treaty negotiations in December of that year.

Once the General Assembly approved negotiations in 1990, the development of an international climate change regime had officially begun.

At Rio and Beyond: The Development of the UNFCCC

At the Earth Summit in Rio De Janeiro in 1992, a number of potential environmental agreements were discussed, and preliminary negotiations took place. One of these agreements was the United Nations Framework Convention on Climate Change (UNFCCC), which was signed by 154 states as well as the predecessor to the European Union, the European Community. The signing of the convention was only the first step in creating the first global climate change treaty. Perhaps because the UNFCCC does not include legally binding commitments, the Framework Convention quickly entered into legal force in March 1994, becoming the first international treaty on global climate change.

Although the UNFCCC does not set binding targets, it "committed the forty industrial countries (listed in Annex I of the agreement), which are largely responsible for creating the problem of climate change, to take the first major steps toward alleviating the problem" (Soroos 2001, 2). The convention explicitly states that each of the so-called Annex I, or industrialized, countries agreed to "adopt national policies and take corresponding measures on the mitigation of climate change, by limiting its anthropogenic emissions of greenhouse gases and protecting and enhancing its greenhouse gas sinks and reservoirs" (United Nations Climate Change Secretariat 1992, article 4.2). More specifically, the signatories to the UNFCCC were expected to implement domestic policies that would reduce their greenhouse gas (GHG) emissions to the 1990 levels. Table 2.2 is a summary of the national commitments made by the countries in the Organisation for Economic Co-operation and Development (OECD) in response to the UNFCCC.

Although all of these countries committed to meeting their voluntary reduction targets, very few signatories to the treaty had actually accomplished these goals by 2001. Table 2.3 is a summary of carbon dioxide emissions for the OECD nations; it ranges from 1990 to 1997, the latter of which being the year that the Kyoto Protocol was written. As can be seen from the table, even with the UNFCCC treaty, overall carbon dioxide emissions in OECD countries had risen 9.7 percent above 1990 levels by 1997 (IEA 2000a).

TOWARD A CLIMATE CHANGE REGIME

Conference of the Parties-1: Berlin and the Berlin Mandate

After the UNFCCC entered into "legal force"—the official term used to describe the international ratification and treaty adoption process—it

Table 2.2. National Target Commitments of Selected OECD Countries in Response to UNFCCC

Country	Stabilization (Target Year)	Reduction (Target Percentage, Year)
Australia	2000	20% by 2005
Austria	—	20% by 2005
Belgium	—	5% by 2000
Canada	2000	—
Denmark	—	20% by 2005
Finland	2000	—
France	2000[a]	—
Germany	—	25–30% by 2005
Greece[b]	—	
Iceland	2000	—
Ireland	—	Limit increase to 20% by 2000
Italy	2000	20% by 2005
Japan	2000[a]	—
Luxembourg	2000	20% by 2005
Netherlands	1995	3–5% by 2000
New Zealand	—	20% by 2000
Norway	2000	—
Portugal[b]	—	—
Spain	—	Limit increase to 25% by 2000
Sweden	2000	2005
Switzerland	2000	2005
Turkey[b]	—	—
United Kingdom	2000	—
United States[c]	2000	—

Source: Adapted from International Energy Agency (1994); Paterson (1996).
Note: Although the European Union had committed to stabilization at 1990 levels by the year 2000 (in response to the UNFCCC), many European countries did not implement national policies.
[a]Stabilization of per capita emissions.
[b]Made no commitments.
[c]Did not make target commitment; instead, made commitment to set policies.

became clear that emission levels were not changing very much. As a result, annual meetings, which are officially known as "Conferences of the Parties," were held to continue international talks on climate change. The Conference of the Parties-1 (COP-1) was held in Berlin in 1995. At COP-1, the nations in attendance agreed that stronger commitments to mitigate climate change were needed. As with most international negotiations, many different opinions about how to proceed were put forward by the parties involved.

On the last day of the COP-1 negotiations, the Berlin Mandate was made public. In summary, the mandate stated that, by 1997, a protocol would be drafted that "contained additional commitments for industrialized countries for the post-2000 period" (Bodansky 2001, 34–35). The

Table 2.3. Changes in Carbon Dioxide Emissions for OECD Countries, 1990–1997
(per million tons)

Country	1990	1997	Change (%)
Australia	259	311	20.1
Austria	59	64	8.5
Belgium	106	119	12.3
Canada	448	479	6.9
Czech Republic	150	128	−17.2
Denmark	51	60	17.6
Finland	53	63	18.9
France	368	350	−4.9
Germany	967	865	−10.5
Greece	70	79	12.9
Hungary	68	58	−14.7
Iceland	2	2	0.0
Ireland	32	36	12.5
Italy	402	416	3.5
Japan	1,049	1,160	10.6
Korea	232	423	82.3
Luxembourg	11	8	−27.3
Mexico	297	339	14.1
Netherlands	157	175	11.5
New Zealand	24	31	29.2
Norway	29	33	13.8
Poland	349	347	0.1
Portugal	40	50	25.0
Slovak Republic	54	32	−40.7
Spain	212	247	16.5
Sweden	52	53	0.2
Switzerland	41	41	0.0
Turkey	138	186	34.8
United Kingdom	572	538	−5.9
USA	4,844	5,468	12.9
Total	11,053	12,130	9.7

Source: IEA (2000a).

mandate responded to the parties' concerns about article 4, paragraphs 2(a) and (b), of the United Nations Framework Convention on Climate Change. In other words, it declared that the parties involved in the negotiations had "agreed that strengthening the commitments by industrialized nations in the Convention on Climate Change was 'the priority' for continuing negotiations, and that in doing so, 'quantified reduction objectives' would be set within 'specified time frames'" (Leggett 1999, 202). As discussed in detail later, the Berlin Mandate stipulated that only the developed countries commit to actual numerical goals, a decision that has had enduring effects on the issue of climate change. As a senior official in the Clinton White House would later put the matter: "The Berlin Mandate has, in my view, been no end of grief" (interview, May 2001).

Conference of the Parties-2:
Geneva and the Geneva Ministerial Declaration

Prior to Conference of the Parties-2 (COP-2), the IPCC released its second assessment report, in December 1995. This report built on the accumulating scientific findings and reached even stronger conclusions regarding the notion that human actions are contributing to climate change. When COP-2 was held, in Geneva in July 1996, the major goal was to agree to legally binding reductions in greenhouse gases. With agreement on this point, the parties would then be prepared to draft a regulatory tool, or treaty, that would be the next step in climate change mitigation. Thus, the text of the Geneva Ministerial Declaration states that the parties instructed their representatives "to accelerate negotiations on the text of a legally binding protocol or another legal instrument to be completed in due time for adoption at the third session of the Conference of the Parties" (UN Climate Change Secretariat 1996). With these words, the treaty that would come out of these years of negotiations was committed to include legally binding and numerical emission reductions that would ultimately lead to the Kyoto Protocol. Agreement on this point was seen as the major goal of this round of negotiations and the key to a successful next round of negotiations the following year, in Kyoto.

Conference of the Parties-3: Kyoto and the Kyoto Protocol

In December 1997, the third round of the Conference of the Parties was held in Kyoto, Japan. The goal of this round of the negotiations was to draft a protocol that would eventually be adopted by all of the parties as a legally binding international climate change treaty. Following the goals of the Berlin Mandate and the Geneva Ministerial Declaration, the Kyoto Protocol included legally binding emission-reduction commitments for the industrialized, or Annex I, countries. The Kyoto Protocol states that "industrialized countries have a legally binding commitment to reduce their collective greenhouse gas emission by at least 5% compared to 1990 levels by the period of 2008–2012."[15] Soroos summarizes the distribution of the emission reductions: "These emission targets vary by country, ranging from an 8% reduction for most European countries and a 7% reduction for the United States, to a freeze for Russia and the Ukraine, and an 8% increase for Australia. These limits are to be achieved as averages for the five years 2008 to 2012, and in most cases are based on 1990 levels" (2001, 2–3).

As with all of the earlier rounds of negotiations, Conference of the Parties-3 (COP-3) involved a significant amount of disagreement. Discussions about the targets, timetable, and wording of the protocol were argued down the last twenty-four hours of the conference; the last day of the conference is said to have included a full twenty-four hours of negotiation without breaks for the delegates to sleep. Much has been written about the contentiousness surrounding the final drafting of the protocol (see, e.g., Flavin 1998; Leggett 1999), but perhaps Flavin best describes the adoption of the Kyoto Protocol, on December 11, 1997. In his words, "With the spotlight of the world's media upon them, delegates decided they had more to fear from a failed agreement than one with which they only partially agreed. . . . Despite remaining reservations, no government was prepared to stand in the way" (1998, 13).

MOVING TOWARD IMPLEMENTATION

Since its approval at COP-3, the Kyoto Protocol has begun the slow progress toward implementation. Prior to the Buenos Aires session of the Conference of Parties (COP-4), the Kyoto Protocol was open for signature. By the middle of 1999, the protocol had been signed by eighty-four countries, including all Annex I nations except Turkey. Signing the protocol is the first step in implementation; afterward, the difficult work of negotiating a ratifiable treaty begins. As has already been pointed out, the Kyoto Protocol includes rules about how it must enter into legal force to become an international environmental treaty. To restate article 25 of the Protocol: "This Protocol shall enter into force on the ninetieth day after the date on which not less than 55 Parties to the Convention, incorporating Annex I Parties which accounted in total for at least 55% of the carbon dioxide emissions for 1990 from that group, have deposited their instruments of ratification, acceptance, approval or accession" (United Nations Climate Change Secretariat 1998, 24).

Conferences of the Parties-4 and 5: Buenos Aires and Bonn

Since 1997, international meetings and Conferences of the Parties have continued to be held to finalize the text of the protocol as well as the mechanisms through which the parties can account for emission reductions. When the Conference of the Parties-4 (COP-4) convened, in Buenos Aires in November 1998, the goal was to design a plan and a timeline to move the Kyoto Protocol toward ratification and adoption. At the meeting, which proved to be relatively uneventful, the so-called Buenos Aires Plan of Action was adopted. "The plan establishes deadlines for finalizing the outstanding details of the Kyoto Protocol so that the agreement will be fully operational when it enters into force sometime after the year 2000."[16] One of the main contributions of the plan was to state the parties' agreement that the Kyoto Protocol would be finished and ready for ratification by 2000. In addition, the parties discussed some of the major mechanisms in the protocol to add specifics to these aspects of the climate change regime.

Like COP-4, the Conference of the Parties-5 (COP-5) round of negotiations, in Bonn in 1999, proved to be somewhat uneventful. The goal of the meeting was to respond to the concerns of the parties and to be certain that the Kyoto Protocol would be ratifiable after the Conference of the Parties-6 took place the following year in The Hague. The major issues discussed at COP-5 involved methods for assessing national emissions, rules for crediting countries for forests (carbon sinks), establishing monitoring of compliance, emissions trading, and procedures for projects in the developing world. These two main issues involved joint implementation (the opportunity for a country to account for some of its emission reductions inside a different country) and the clean development mechanism (the opportunity for industrialized countries to account for some of their emission reductions through projects in developing countries).

Conference of the Parties-6, 7, and 8: The Hague, and Bonn

In contrast to the two meetings before it, the COP-6 round of negotiations, in 2000 in The Hague, was not exactly uneventful. Per the agreement made in Buenos Aires, the COP-6 round was expected to conclude discussions of the text of the Kyoto Protocol and yield a ratifiable treaty. Despite the high hopes of the parties, however, no such agreement was achieved. In fact, on the second-to-last day of the negotiations, the talks broke down, with the delegation from the European Union walking out. Discussions between the parties resumed but not in time to tie up the loose ends that remained in the Kyoto Protocol. In response to the failed meeting, the president of the Conference of the Parties, Jan Pronk, the environment minister from the Netherlands, proposed that the parties reconvene in the near future to try to resolve the remaining issues. After much dialogue about where and when the talks could be held, they were set to resume in July 2001 in Bonn, under the title of the Conference of the Parties-6bis (COP-6bis). Although papers have been published presenting opinions about the breakdown in the talks (see, e.g., Müller 2000; Paterson 2001; Vrolijk 2001), there is very little agreement about the cause of the failure of The Hague round. In fact, during the final hours of the talks, I witnessed environmental groups, who were involved in the negotiations as nongovernmental organization (NGO) observers and who were members of the same

Climate Action Network, present two separate statements to the press—one of which blamed the United States and the other of which blamed the European Union for the failure of the negotiations (Environmental Defense 2000; Friends of the Earth 2000). Clearly, something went seriously wrong at the end of the COP-6 negotiations in The Hague. Equally clear was that not even environmental groups could agree on whom to blame. Even with these inauspicious events, coupled with the Bush administration's withdrawal from the negotiations of the United States in March 2001, the remaining parties to the Kyoto Protocol moved forward on finalizing the text at the COP-6bis negotiations the following summer.

This book follows the national and international debates as they unfolded, leading up to the conclusion of the COP-6bis round of negotiations and the movement of most countries to ratify the Kyoto Protocol. The negotiations following the COP-6bis, and the attempt to enter the protocol into legal force, are discussed at the conclusion, chapter 7. Because the goal of this book is to examine the divergent responses of states to the Kyoto Protocol, it is important to do so in a way that is theoretically informed. As such, the next chapter reviews the ways that scholars have explained the relationship between society and the natural environment.

NOTES

1. See, for example, Chambers (n.d.); United Nations University (1999).
2. See, for example, OECD (1997, 1999a, 1999b, 1999c).
3. See, for example, IEA (1994, 2000a, 2000b, 2000c).
4. See, for example, Nichols and Martinot (2000).
5. See www.birkie.org/races/history.html (accessed October 2, 2003).
6. See www.unfccc.int/resource/iuckit/fact17.html (accessed October 1, 2003).
7. See www.ipcc-nggip.iges.or.jp (accessed October 1, 2003).
8. See, for example, Andersson and Gupta (1998); Blowers and Glasbergen (1996); Choucri (1993); Downs, Danish, and Barsoom (2000); Haas (1989, 1990, 1995, 2000); Haas and Sundgren (1993); Luterbacher and Sprinz (2001); Porter and Brown (1991); Rowlands and Greene (1992); Tolba and Rummel-Bulska (1998); Young (1989, 1994, 1997, 1998); Young, Demko, and Ramakrishna (1996).
9. See, for example, Kawashima (1997); Messner and others (1992); Paterson (1996, 2001); Rabe (1999); Schreurs and Economy (1997); Soroos (2001); Stone (1992); Sugiyama and Michaelowa (2000); Young (1994, chapter 2).
10. See, for example, Boehmer-Christiansen (1997); Freudenburg and Buttel (1997); McCright and Dunlap (2000); Lutzenhiser (2001); Paterson (1996, 2001); Rayner and Malone (1998); Soroos (2001).
11. See, for example, Anderson (2001); Dunn (1998); Flavin (1998); Judis (1999); Sarewitz and Pielke (2000).

12. Scholars working on the environmental "Kuznets curve" find that the relationship between some pollutants and affluence follow an inverse, U-shaped curve. In other words, as affluence increases, pollution levels initially increase; however, once a certain level of affluence is attained, pollution levels decrease.

13. See, for example, Bord, Fisher, and O'Conner (1998); Bostrom and others (1994); Brechin (2003); Dunlap (1998); Read, Bostrom, and Granger-Morgan (1994); von Storch and Bray (1997); see also Meijnders (1998).

14. Although table 1.1 provides summaries of events through 2002, this chapter only discusses the events through the Conference of the Parties-6 meeting in The Hague in 2000. The rest of the events are discussed later in the book.

15. See www.unfccc.int/resource/iuckit/fact17.html (accessed October 1, 2003).

16. See unfccc.int/cop4/infomed/p111498.html (accessed October 2, 2003).

3

Empirically Analyzing the Material Characteristics of the Environmental State and Moving toward Understanding the Political Characteristics

As chapter 2 indicates, nation-states have been working on the formation of a global climate change regime for many years. Although the issue itself has been around since the 1970s, the politics of the issue have yet to be resolved in the global arena. Given the ongoing disagreements about how the world should move beyond the UNFCCC to regulate this global environmental issue, the case of climate change provides an opportunity to resolve some of the differences within the sociological literature regarding the relationship between economic growth and environmental degradation. As pointed out in chapter 1, the differences between the "environmental state" and the "environmental sociology" lines of thought can be traced largely to their starkly differing expectations about the nature of the relationship between state regulation and the economy. Much of the recent work on the environmental state reflects the view that environmental protection is, in the words of Anthony Giddens, "a source of economic growth rather than its opposite" (Giddens 1998, 19). In contrast, many scholars who contribute to the environmental sociology literature come to nearly opposite conclusions.

There are two approaches for resolving this difference in views—a broader, quantitative approach; and a deeper, qualitative case study approach. This chapter follows the first, or more quantitative, of the approaches, analyzing the relationships between the characteristics of advanced nations and their contributions to climate change. As will be seen, however, the results of this analysis raise additional questions regarding the relationship between postindustrial states and the environment. Thus, I follow by presenting a framework for going beyond the quantitative analysis. I look more deeply at select developed nations using a qualitative case study method that expounds on the dynamics of the global environmental system. Finally, I describe the methodology through which I analyze the role of social actors—from industry, civil society, science, and the state—in forming the disparate climate change regimes we see today.

MEASURING MATERIAL OUTCOMES

The issue of global climate change provides an opportunity to look at an "actual environmental protection outcome" (Buttel 2000a, 188) while the issue is still being discussed in both the science world and the policy world. Perhaps the environmental outcome related to global climate change that has received the greatest amount of attention in international circles to date has involved national-level emissions of carbon dioxide, or CO_2. This compound is the largest single constituent of so-called greenhouse gases—the emissions that are now considered by the majority of atmospheric scientists as contributing to global warming by increasing the earth's propensity to retain the sun's heat (see, e.g., IPCC WGI 2001; National Research Council 1992, 2001). As noted by Roberts and Grimes, "Carbon dioxide is now understood to account for over half of the effect of greenhouse warming" (1997, 192; see also Dietz and Rosa 1997).

Although recent research from institutions such as the OECD has begun to look at the "other greenhouse gases" (see, e.g., Burniaux 2000), there are other valid methodological reasons for focusing on carbon dioxide. As noted by Dietz and Rosa (1997), "Data on other greenhouse gases are also less reliable than the industrial CO_2 data. Current estimates of CH_4 (methane) emissions are uncertain to at least a factor of two and do not take account of biomass burning, which may contribute perhaps one-fifth of the total anthropogenic emissions. Data on chlorofluorocarbons are reported as an aggregate for the European community nations, which are among the highest chlorofluorocarbon producers and consumers, and nitrous oxide emissions are available only for a handful of nations" (Dietz and Rosa 1997, 77). In addition, it is carbon dioxide emissions that determine whether the Kyoto Protocol can be entered into legal force and become an actual treaty.

In other words, in the views of the relevant scientists and many policymakers, there is widespread agreement on the need for reducing CO_2 emissions as required under the provisions of the Kyoto Protocol in the collective best interest of humanity. At the same time, those measures may well be resisted quite intensely by key economic actors within many of the individual nation-states, a topic that is discussed in more detail later.

Data and Methods

Given that there is little reason to expect environmental state theories of "advanced," or "late," capitalistic development to apply to nations that have only limited economic prosperity—and given that just thirty developed countries emit over half of the world's carbon dioxide (IEA 2001)[1]—I limit my examination of CO_2 emissions to the more prosperous, or more developed, nations of the world. Specifically, my analysis focuses on twenty-nine out of the thirty nations that belong to the Organisation for Economic Co-operation and Development, or OECD. (The one excluded nation is the most recent addition to the OECD, the Slovak Republic, for which few data are yet available).

Given that these twenty-nine nations differ greatly in terms of their population sizes, the dependent variable for "environmental protection outcomes" is standardized by population numbers—that is, I focus on CO_2 emissions per capita. The International Energy Agency (IEA), an autonomous agency linked with the OECD that concentrates on energy issues, compiles and maintains data on the energy consumption and production for most of the countries in the world.[2] Included in the data that they track is a measure of total CO_2 emissions per capita for most countries in the world (in tonnes of carbon dioxide per person per year). The most recent CO_2 emission inventory provides data from 1998 (IEA 2001).

To address the two differing ways in which the recent environmental state theorists and the earlier environmental sociologists have dealt with the relationships between prosperity and environmental quality, I bring in two sets of independent variables. The first set involves economic indicators, and the second includes a collection of environmental performance indicators. Also, in response to suggestions from colleagues who have reviewed earlier drafts of this chapter, I include other types of independent variables: three measures of what I call "environmental institutionalization" and the two best-known assessments of national environmental protection performance that have been produced by independent organizations. Table 3.1 presents a list of the variables used and their data sources.

The first set of independent variables, involving economic indicators, reflects the perspective of environmental sociology theorists as well as many present-day representatives of potentially regulated industries, that economic prosperity is associated with environmental degradation—or, to be more specific, that emissions are generally proportionate to the size of the economy. As will be recalled, these expectations are quite different from the views put forth by Giddens (1998, 19), in which "environmental

Table 3.1. Quantitative Variables and Data Sources

Category	Variable Name
Dependent variable	CO_2 emissions per capita (tonnes of CO_2 per person 1998)[a]
Economic measures	GDP per capita (billion U.S.$ per person 1998)[b]
	Total primary energy supply per capita (MTOE per person 1998)[a]
Ecological efficiency measures	Municipal waste (kg per person 1998)[b]
	Industrial waste (kg per U.S.$1,000 GDP 1998)[b]
	% change in total 1980–1997 primary energy supply (MTOE)[b]
	Motor vehicle travel per capita (billion vehicle-km 1997)[b]
Environmental institutionalization measures	National parks and protected areas (percentage of total land area)[c]
	Country chapters of international environmental nongovernmental associations (annual number)[d]
	Nation-state contributions to intergovernmental environmental organizations ($U.S. contributed/GDP)[e]
Independent measures of environmental impact	Ecological footprint (1997)[f]
	Sustainability index (2001)[g]

[a]IEA (2001)
[b]OECD (1999c)
[c]World Commission on Protected Areas (2001)
[d]Union of International Associations (2000)
[e]Stokke and Thomessen (2001)
[f]Wackernagel and colleagues (1997)
[g]Global Leaders of Tomorrow Environment Task Force (2001)

protection is seen as a source of economic growth rather than its opposite." To test the competing expectations straightforwardly, I draw on two economic measures. Both are considered by the OECD to be "selected economic indicators," although one is well known but potentially biased against measures of actual environmental outcomes, while the other is less well known but may have fewer such biases. The first measure involves each nation's gross domestic product (GDP) per capita, as of 1998 (OECD 1999c). Although this measure is well known and widely used, it has been criticized by a number of economists[3] as being an excessively "gross" measure, in that it merely adds up a nation's economic transactions. A nation's GDP goes up, for example, when workers receive wages for cutting down a forest—and it goes up even more if the resultant deforestation leads to other expenditures (e.g., for rebuilding homes that are destroyed in downstream flooding, or even for burying flood victims or hiring lawyers to sue the logging company). If a forest is not cut down but instead remains standing, then—even though it can help to mitigate global climate change by absorbing carbon dioxide—this social benefit is not reflected in GDP figures unless direct monetary transactions take place.

The lesser well-known economic indicator involves each nation's total primary energy supply (TPES) per capita, measured in million tonnes of oil equivalent (MTOE), again standardized by population (IEA 2001). This indicator is the most straightforward of any of the available measures of the ways in which a nation contributes to CO_2 emissions through its energy consumption.[4] As recent analyses have shown (Hale 1997; Roberts and Grimes 1997; see also, Farla and Blok 2000), there are strong correlations between energy consumption and GDP, but increases in the energy efficiency of most national economies have led to a substantial "decoupling" of energy inputs and economic outcomes over the past three decades, particularly among the most prosperous nations of the globe. Similarly, total energy consumption is "decoupled" from CO_2 emissions because available energy technologies are not equal in the amount of CO_2 they produce.

The second set of variables reflects the expectations of scholars such as Inglehart (1995) or Spaargaren and Mol (1992), namely, that there may well be national-level differences in readiness to take tangible steps to protect the environment. One possibility is that those nations with the greatest willingness to curb CO_2 emissions may also be the ones that would be taking steps to ensure other types of positive environmental protection outcomes. As no standard indicators exist, I have included four possible measures of what I call "ecological efficiency," the measures of which have all been compiled by the OECD. The first of which offers the most straightforward measure available of the extent to which individual consumers have become part of a so-called throwaway culture, and it involves the number of kilograms of municipal waste discarded per capita in 1998. By contrast, the second is better understood as a measure of the wastefulness (or conversely, the efficiency) of each nation's industries: kilograms of industrial waste produced per U.S.$1,000 GDP, also for 1998 (except in the cases of Canada and the United States, where no data were available for 1998, and where I have instead used the most recent data available, from 1994 [OECD 1994]). This second measure does have an important potential weakness that needs to be noted with respect to the present analysis, namely, the fact that it is not standardized in terms of population. In contrast to most of the other relevant variables, this measure is standardized in terms of monetary production output; however, this form of standardization is generally considered by the OECD and others to provide a better measure of a nation's industrial or production efficiency, since industrial output may or may not be associated with a nation's population size.

My third indicator of ecological efficiency is the percentage change in energy consumption. It is measured by the change in the total primary energy supply (TPES) from 1980 to 1997, and it is coded so that positive numbers indicate increases in energy consumption and negative numbers indicate decreases. Inclusion of this variable in the analysis responds to the emphasis that has been placed on relatively recent improvements in energy efficiency, particularly after the energy price shocks of the mid- to late 1970s (Roberts and Grimes 1997; and Hale 1997). The fourth such variable measures kilometers of motor vehicle travel per capita in 1997 (OECD). This variable has a good deal to do with CO_2 emissions, in that 22.7 percent of emissions in OECD countries in 1997 came from the transport sector (IEA 2000b). In addition, this variable provides an indicator of the efficiency of the transportation infrastructures in OECD nations, in that motor vehicle travel reflects national policy choices—including the willingness to invest in well-developed rail systems, pay higher gasoline taxes, and plan and build higher-density urban areas that are likely to have fewer emissions.

The third set of independent variables are similar to Frank, Hironaka, and Schofer's indicators of "environmentalization . . . the global institutionalization of the principle that nation-states bear responsibility for environmental protection" (2000a, 96). These variables reflect top-down actions by nation-states—or what I call environmental institutionalization—to protect the environment. The first measure is the same as that used by Frank, Hironaka, and Schofer (2000a), namely, the number of chapters of international environmental nongovernmental associations in each country (Union of International Associations 2000). The second variable reflects a country's commitment to its membership in international environmental organizations. This variable is the percentage of a country's GDP in 1998 that was contributed to three of the largest funds for international environmental organizations: the Multilateral Fund for the Implementation of the Montreal Protocol, the Global Environmental Facility, and the Environment Fund of the United Nations Environment Programme (Stokke and Thomessen 2001). Like the industrial waste measure, this variable is standardized monetarily per production output as a way of gauging a country's monetary support of international environmental organizations. The third variable measures the amount of land that has been set aside as a national park or a protected area, as a percentage of the area of the country. These numbers are compiled by the United Nations Environment Programme's World Conservation Monitoring Center Protected Areas Database, for the year 2001.[5]

For my fourth set of independent variables, I make use of the two best-known efforts by other researchers to produce composite measures of nations' overall environmental impacts and environmental quality. The first is the "ecological footprint of nations" measure, developed by Wackernagel and colleagues to quantify ecological impacts (1997; see also Wackernagel and Rees 1996).

The basic intent of the footprint variable is to measure each nation's resource consumption and waste accumulation, relative to its productive land area. This variable has been praised by York, Rosa, and Dietz (2003, 280) as "the most comprehensive measure of environmental performance available" (see also Wilson 2001; Wright and Lund 2000), although other academics have criticized the measure for ignoring the role of trade (see, e.g., Ayres 2000). The second such measure is the environmental sustainability index, which was developed by scholars at the Yale University Center for Environmental Law Policy, the Earth Institute Center for International Earth Science Information Network (CIESIN) at Columbia University, and the Global Leaders for Tomorrow Environment Task Force of the World Economic Forum (Global Leaders of Tomorrow Environment Task Force 2001). The sustainability index was developed to represent "a country's environmental success . . . in the management and improvement of common environmental problems"[6] and to measure a nation's "overall progress toward environmental sustainability."[7] Various organizations and academics, however, have criticized this index with the New Economics Foundation, for example, calling it a measure of "global misleadership" (Capella 2001). Rather than siding arbitrarily with past assessments that have argued either for or against these measures, I plan to include each of the measures in the analyses as a way of testing the empirical utility of the available approaches.

Bivariate Analysis

Given the exploratory nature of the analysis, the first step was to perform a screening of all of the independent variables at the level of basic construct validity. The first question posed was: At least at the zero-order level, were the potential explanatory variables correlated with CO_2 emissions in the expected direction? As can be seen from table 3.2, most of the potential explanatory variables pass this relatively simple test, but one variable clearly fails—the second of the overall environmental performance indicators, namely, the so-called sustainability index. As can be seen from the positive correlation for this index, the nations having higher scores—meaning the ones that were supposedly the more "sustainable"—actually had higher levels of CO_2 emissions than those that were identified as being less "sustainable." Because of the failure to pass this simple test of validity, this measure was dropped from further analysis. All other variables were retained for subsequent steps, but it is worth drawing attention to three variables that fall into an intermediate category. Each of the three variables that measures environmental institutionalization is also positively associated with CO_2 emissions. Given the importance of these variables in addressing questions raised by Buttel (2000a) regarding the relationship between environmental

Table 3.2.   Product-Moment Bivariate Correlation Coefficients for OECD Countries'
CO_2 Emissions per Capita (N = 29 countries)

	Correlation Coefficient
GDP per capita (1998)	.648**
Total primary energy supply per capita (1998)	.677**
Municipal waste (1998)	.484**
Industrial waste (1998)	.519**
% total primary energy change (1980–1997)	−.260
Motor vehicle travel per capita (1997)	.723**
National parks and protected areas	.281
Country chapters of international environmental nongovernmental associations	.300
Nation-state contributions to intergovernmental environmental organizations	.214
Ecological footprint	.518**
Sustainability index	.297

*$r < .05$
**$r < .01$

bureaucratization and actual environmental outcomes—and also given the fact that multivariate analyses could still ultimately suggest that these positive zero-order correlations are due to other factors—these three variables clearly need to be retained in the analyses; all the same, the fact that all three would be positively associated with CO_2 emissions, even at the zero-order level, is a potential cause for concern and a reason for closer attention.

Beyond the failure of the environmental sustainability index to meet the validity test, these bivariate correlations indicate that a number of the variables may significantly contribute to a nation-state's levels of CO_2 emissions per capita. In fact, the variables that are significantly associated with the dependent variable do not fit clearly into those that represent characteristics of the earlier environmental sociological theories, nor do they reflect the environmental state literature. First, as might be expected on the basis of work by Roberts and Grimes (1997) or Dietz and Rosa (1997), both of the independent variables that have been included in the analysis to reflect expectations of the earlier environmental sociologists—GDP per capita and total primary energy supply per capita—are significantly associated with CO_2 emissions per capita (r = .648, .677, respectively). These results reflect the tendency for the more prosperous countries to contribute more to global climate change. At the same time, however, three of the four variables that represent characteristics of the environmental state literature—municipal waste per capita, industrial waste per unit of economic output, and motor vehicle travel per capita—are also significantly associated with CO_2 emissions per capita (r = .484, .519, .723, respectively), as is the variable that signifies each nation's ecological footprint (r = .518). These latter findings suggest that the countries that are less ecologically efficient tend to contribute more to global climate change, while only the 1980–1997 change in energy consumption fails to be significantly correlated with CO_2 emissions per capita. The more telling question, however, has to do with the influence of each variable once other independent variables are controlled.

Multiple Regression Analysis

Accordingly, my next step was to move to multivariate analysis of the relationship between the independent variables and the dependent measure of CO_2 emissions per capita. The most straightforward approach to such an analysis is through the use of ordinary least squares (OLS) regression, but certain safeguards are important when engaging in such forms of analysis, particularly given the obvious potential for multicollinearity that exists when using national-level data. In the analysis that follows, I employ three relatively standard safeguards (in addition to the face validity examination summarized earlier) and a fourth and final double check.

The first such safeguard is the most formal, involving the explicit consideration of tolerance statistics. The standard rule of thumb is to exclude a variable from the analysis if its tolerance level drops below 0.01, or if it causes the tolerance of other variables to drop below that same level; in the present case, this test indicates no problems of multicollinearity that are severe enough to create reasons for concern.

The second involves the practice of backward elimination, a process that involves "simplifying a regression by dropping non-significant variables" (Hamilton 1990, 581–82). In the following analysis, nonsignificant variables were dropped from the analysis one at a time (per Hamilton), beginning with those variables that were furthest from achieving statistical significance and continuing until all remaining variables exceeded standard levels of statistical significance ($p < .05$).

The third safeguard follows directly on the second, and it involves both paying attention to the coefficients that remain in the equation and being on the alert for wild fluctuations. Although GDP per capita changes signs when the ecological footprint variable is removed from the analysis, the variable is not significant and does not become significant when reintroduced in the final equation. The results of this process are presented in table 3.3, which summarizes the results from the analyses, allowing readers to assure themselves that no other such wild fluctuations emerge.

In the interest of providing readers with information on all of the initial statistical relationships, the full model (the key findings from which table 3.3 derives) includes all of the variables, although these results need to be interpreted with caution, in light of the potential for multicollinearity noted earlier. As can be seen from table 3.3, only one of the independent

Table 3.3. Standardized Regression Coefficients (and unstandardized regression coefficients) and Significance Level for Regression of Carbon Dioxide Emissions per Capita on Selected Independent Variables (OECD 1990; $N = 29$)

Independent Variable	Full Model	Second Model	Third Model	Fourth Model	Fifth Model
GDP per capita 1998	−.004 (.307)	(−.03004)−			
	.992				
Total primary energy supply per capita 1998	.161 (.307)	.161 (.305)	.153 (.290)	.130 (.253)	.122 (.236)
	.482	.454	.456	.436	.452
Municipal waste 1998	.163 (.005396)	.162 (.005359)	.160 (.005279)	.119 (.004228)	.120 (.004257)
	.424	.340	.332	.434	.421
Industrial waste 1998	.425 (.0227)**	.425 (.02275)**	.428 (.02287)**	.475 (.02662)**	.483 (.02705)**
	.006	.005	.003	.001**	.000
% change in energy consumption 1980–1997	.097 (.005727)	.096 (.005709)	.098 (.005824)	.101 (.00338)	.110 (.006914)
	.483	.461	.439	.420	.358
Motor vehicle travel per capita 1997	.684 (.840)	.682 (.838)*	.656 (.805)**	.547 (.707)*	.540 (.699)**
	.061	.021	.009	.012	.000
National parks and protected areas	.091 (.04125)	.091 (.04120)	.090 (.04091)	.058 (.02180)	.053 (.02562)
	.491	.477	.468	.624	.645
Country chapters of international environmental nongovernmental associations	−.030 (−.04321)	−.030 (−.004327)			
	.854	.849			
Nation-state contributions to intergovernmental environmental organizations	−.170 (−.000006)	−.171 (−.000006060)	−.171 (−.000006062)	−.041 (−.000001363)	
	.343	.207	.194	.737	
Ecological footprint 1997	−.165 (−.304)	−.165 (−.305)	−.140 (−.258)		
	.533	.520	.512		
Constant	(−.298)	(−.300)	(−.336)	(−1.029)	(−1.118)
Adjusted R^2	.653	.672	.689	.692	.705

Note: Unstandardized regression coefficients are in parentheses; significance levels are italicized.
*p < .05
**p < .01

Independent Variable	Sixth Model	Seventh Model	Eighth Model	Final Model
GDP per capita 1998	—	—	—	—
Total primary energy supply per capita 1998	.111 (.216) *.479*	—	—	—
Municipal waste 1998	.124 (.004383) *.399*	.133 (.004697) *.359*	—	—
Industrial waste 1998	.485 (.02713)** *.000*	.511 (.02861)** *.000*	.488 (.02734)** *.000*	.466 (.02610)** *.000*
% change in energy consumption 1980–1997	.102 (.006394) *.380*	.118 (.007430) *.294*	.122 (.007665) *.278*	—
Motor vehicle travel per capita 1997	.561 (.726)** *.006*	.639 (.827)** *.000*	.733 (.949)** *.000*	.688 (.891)** *.000*
National parks and protected areas	—	—	—	—
Country chapters of international environmental nongovernmental associations	—	—	—	—
Nation-state contributions to intergovernmental environmental organizations	—	—	—	—
Ecological footprint 1997	—	—	—	—
Constant	(−.941)	(−.957)	(.536)	(1.341)
Adjusted R^2	.715	.721	.722	.719

Note: Unstandardized regression coefficients are in parentheses; significance levels are italicized.

*p < .05

**p < .01

variables in the full model—kilograms of industrial waste produced per U.S.$1,000 GDP—is significant at the level of $p < .05$ when all the remaining independent variables are included in a multivariate analysis. In addition, one additional variable—vehicle travel per capita—is nearly significant in the full model ($p < .10$). The middle columns of this table show the coefficients of the variables that remain in the equation after each of the "least significant" variables is eliminated, one at a time. The final regression equation yields an adjusted R-squared of .719, with the two significant predictors of CO_2 emissions being per capita motor vehicle travel ($\beta = .688$) and per dollar industrial waste generation ($\beta = .466$).[8]

The last of the double checks involves a final pair of tests regarding the differing expectations derived from the work in environmental sociology and the environmental state. For the environmental sociology literature, and for the argument regarding the "enduring conflict" between the economy and environmental protection (Schnaiberg and Gould 1994), it is important to make sure that the dropping of per capita gross domestic product is not merely an artifact of the way in which the analysis has been carried out. It is not. Even when I attempt to force this variable back into the final equation, along with motor vehicle travel and industrial waste generation, the GDP measure proves to have no statistically significant effects on CO_2 emissions per capita ($p > .300$). Similarly, for the environmental state literature, there are no statistically significant effectswhen I attempt to add back in each of the three measures of environmental institutionalization adapted from Frank, Hironaka, and Schofer (2000a)—national parks/protected areas, chapters of environmental NGOs, and contributions to intergovernmental environmental organizations; ($p > .600$) in combination with motor vehicle travel and industrial waste generation.

Discussion of the Material Outcomes

In contrast to the usual expectations of environmental sociologists of past decades, straightforward economic indicators such as the gross domestic product proved not to have significant effects on CO_2 emissions in any of the multivariate analyses. Despite the apparent strength of the convictions of relevant U.S. political leaders (discussed in detail in chapter 6), the correlations between CO_2 emissions per capita and economic output per capita drop to insignificance once other variables are controlled.

Instead, the two measures that jointly explain nearly three-fourths of the variance across industrialized nations in CO_2 emissions per capita have far more to do with policy choices: the degree to which a nation's transportation infrastructure has become dependent on individual vehicles, as measured by the number of vehicle-kilometers of travel per capita; and the ecological efficiency of national industrial output, as measured by non-CO_2 waste from manufacturing industries, per unit of economic output. In short, the "actual environmental outcomes" of CO_2 emission levels are predicted not by overall prosperity levels but by two straightforward measures of the ecological efficiency of national economies—a finding that is all the more remarkable in that the measure of industrial emissions is standardized not in terms of population (as is the case for CO_2 emissions) but in terms of economic output. It is also important to note that the measures of environmental institutionalization fall out of the equation.

At the same time, however, these results cannot be said to provide clear support for the newer environmental state theories. Rather than show international convergence toward the actual acceptance of environmental protection as a "basic state responsibility," the findings merely suggest that it does appear possible for such a state to emerge. In other words, these results suggest that industrialized states do indeed have the potential to shape economic choices—doing so in ways that can clearly affect material outcomes, including those that have an effect on the natural environment. To date, the actual levels of constraints appear to vary widely across the nation-states of the Organization for Economic Co-operation and Development.

MEASURING POLITICAL OUTCOMES

If developed states can control emissions (but only some of them do actually control their emissions), the question that remains is: What explains that variation? It may be significant that much of the debate surrounding the political regulation of global climate change has revolved around nuanced interpretations of the issue, while research on society–environment relationships has been carried out at relatively high levels of abstraction, without including an explicit focus on empirical evidence. Another factor that may well be related involves the fact that so many of these sociological debates have been carried out in terms of relatively stark either–or, black–white distinctions. Such debates, of course, are important, and in view of the complex and multifaceted issues involved, they are certainly deserving of continued theoretical examination. At the same time, however, the existing literatures on environmental sociology and the environmental state do not provide as much guidance toward understanding the complex issues surrounding the potential mitigation of global climate change as one might like. In fact, as I point out in chapter 1, these theories tend to overlook the interrelations among social actors at the domestic and international levels that determine such material outcomes. To gain a deeper understanding of the mechanisms that contribute to the variable responses of nation-states to the potential regulation of climate change, I chose to do a much more detailed, firsthand qualitative examination of a few of the countries that are determining the political and material outcomes of the global climate change regime.

Although the quantitative analysis summarized here has helped us to understand better the contributions of nation-states to global climate change, we are still left with more questions than answers. In particular, these analyses do not help us to understand the conditions under which advanced, or postindustrial, states will see the kinds of outcomes predicted by theorists of the environmental state, nor do they explain why certain nation-states included in the aforementioned analysis have been politically supportive of an international treaty for global climate change while others have not.

An obvious next step in this research is to look more closely at the differing domestic responses to the Kyoto Protocol, during the time when the actors were debating the issues inside their respective countries and then presenting their positions at international negotiations. This analysis not only deepens our understanding of the variable responses of nation-states to the possible regulation of greenhouse gases, it also provides an opportunity to explore the specific characteristics of the global environmental system. Thus, the next step in this research requires the shifting of the level of analysis to the specific social dynamics within nation-states.

Methodology: Selecting Case Study Sites and Delineating Independent Variables

Even though the results of quantitative analysis show that the ecological efficiency of nation-states are the most important in explaining the variation between the emission levels of the OECD nations, they do not fully explain the social dynamics inside the countries—including the very policy dynamics that shape the types of travel available to citizens of a country and the amount of waste a country's industries produce.

This comparative case study design was crafted to facilitate comparison of the influence of various social actors on the domestic responses to the Kyoto Protocol. The first two cases, the United States and Japan, are not only the world's top two economies but are also the world's top two greenhouse gas emitters. Together, these countries represent over 28 percent of the global carbon dioxide emitted in 1998. The ideal third case would be the European Union, as it is the third global leader involved in the negotiations regarding climate change and emits almost as much carbon dioxide per year as the United States.[9] Laws in the European Union (EU), however, continue to be passed by the parliaments of its member states. Thus, the third case must be a country within the EU, which will make comparing the three cases easier.

For the third case study, I have chosen the Netherlands, a decision I made based on three reasons. First, the Netherlands held the presidency of the EU during the climate change negotiations in Kyoto, where the protocol was drafted; second, as has been stated by Gummer and Moreland (2000, 28), the Netherlands has been "considered an environmental leader both within the EU and internationally"; and third, given the fact that much of the work on ecological modernization was developed in the Netherlands, this nation-state provides a particularly instructive case for understanding the EU position. Thus, by including countries that represent the largest political actors on the issue, as well as ones that provide a range of expected fits to the environmental state literature, I hope to be able to answer the theoretical and empirical questions raised in this book.

In contrast to the preliminary quantitative analysis that measured emission levels in postindustrial nation-states, the two key dependent variables for the qualitative analysis address the political side of climate change: first, the political engagement of a country in the international climate change regime; and, second, the formation of a domestic climate change regime that reduces national greenhouse gas emissions. Although a nation-state's intention is sometimes difficult to determine, the first variable is an indicator of a stated commitment to the regime, as measured by a country's expressed intention to ratify the Kyoto Protocol. The second measure, the formation of a domestic climate change regime, is operationalized by the change in national carbon dioxide emissions since the Kyoto Protocol was written, providing a measure of the efficacy of a nation's commitment to reducing emissions. In other words, this variable is meant to determine how well the stated climate change regime has been working, helping us to understand whether the intentions of the nation-state are merely symbolic and allowing us to understand better the nature of the political outcome itself by looking at material outcomes. Given the timeline of the Kyoto Protocol's first commitment period, a nation-state must have begun to reduce its emissions by 2000 to meet the reduction commitments stipulated by the protocol.

The four independent variables in the research are reflective of the four main social actors that have been identified by the environmental state literature and elsewhere, as interacting to shape each country's domestic policy on climate change. By way of introduction, each of these variables is discussed briefly.

The State

The role of the state in society has been one of the main themes within mainstream sociology. In recent years, much of this research has looked at the level of autonomy that a state has in relation to other social actors. Of particular interest is the discussion surrounding what Skocpol calls the

"Tocquevillian" approach (1985). In her own words, this perspective claims that "states matter not simply because of the goal-oriented activities of state officials. They matter because their organizational configurations, along with their overall patterns of activity, affect political culture, encourage some kinds of group formation and collective political actions (but not others), and make possible the raising of certain political issues (but not others)." In this discussion, Skocpol proceeds to describe states as being strong or weak depending on their independence and effectiveness in implementing official actions. Following in this vein, I classify the states within this study in terms of their relative strength or weakness in unilaterally implementing climate change policies (for a more detailed discussion of the literature on the state see, e.g., Poulantzas 1978; Skocpol 1979; Nordlinger 1981; Evans, Rueschemeyer, and Skocpol 1985; Block 1987; Evans 1995).

The Market

Some of the most relevant work about industry's role in relation to other social actors has been produced by scholars working to understand the interactions between the state and the market. Perhaps one of the most useful studies is provided by Mills, in his seminal work on the power elite (1959). Although this work also addresses issues of the state, one of its main themes involves the dominance and effectiveness of business interests in their workings with policymakers (see also Miliband 1969; Domhoff 1990). In the words of Levy and Eagan, these types of relationships are "accomplished through a dense network of relationships between business and the state, including membership in political or social organizations, the 'revolving door' of decision-making personnel between business and the state, politicians' dependence on private donations to fund election campaigns, and business organized and supported think tanks and policy organizations" (1998, 3; see also Galanter 1974; for a more detailed discussion of the scholarship on corporatism and the state see Grant 1985; Cox and O'Sullivan 1988). As an attempt to simplify the complex relationships between the market and other social actors, I classify the role of industry in each case as being either autonomous or collaborative in its responses to the issue of climate change.

Civil Society

The notion of civil society is also a central concept within contemporary sociology (see, e.g., Dewey 1927; Gramsci 1971; Habermas 1989, 1998; Calhoun 1992). Within this vast literature, the notion and roles of civil society are derived and discussed. Much of the literature focuses on conceptualizing the evolving role of the citizen in society. To date, civil society continues to be seen as a social sphere that is separate "from both state and economy" (Cohen and Arato 1994, ix). Not only do civil society actors continue to be considered distinct and separate from other social spheres, but civil society is viewed as consisting of a separate and distinct institutional framework. In fact, Emirbayer and Sheller "contend that the state, economy and civil society are realms of social life whose relative independence from one another constitutes one of the principal hallmarks of modernity. Many of the dynamics of contemporary society are captured in the relations among these empirically interpenetrating and yet analytically distinct institutional domains. . . . [Civil society is] the institutional sector that, metaphorically speaking, lies 'in between' the state and economy" (1999, 151). In short, civil society is seen as its own "institutional complex" (Emirbayer and Sheller 1999, 152). Inside this complex, or sphere, lie social movement organizations, civic associations, transnational nongovernmental organizations, and citizens who voice their political preferences through their demonstrations, votes, and pocketbooks (for a more detailed discussion of the literature on civil society, see Cohen and Arato 1994; Emirbayer and Sheller 1999; see also Hann and Dunn 1996). Perhaps in its most general form, civil society has come to be defined as involving a "self-organized citizenry" (Emirbayer and Sheller 1999, 146; for a complete discussion, see Cohen and Arato, 1994; see also, Hann and Dunn, 1996). Within this study, I classify civil society as being either internal or external to the policy-making process.

Science

Research into environmental issues and environmental regimes has looked extensively at the role of science in policy making at the national and international levels (see, e.g., Andresen et al. 2000; Barrett, Chambers, and Schroeder 2001; Jasanoff and Wynne 1998). Perhaps Andresen and colleagues best summarize the issues surrounding this research: "The transformation of research-based knowledge into policy premises hinges not only upon the formal institutional setting in which it occurs, but also upon the skills and behavior of individuals occupying important boundary roles" (2000). Of particular interest to this study is the relationship between science and the domestic policy-making process in each country. Thus, science is characterized as being either peripheral or central to the policy decisions surrounding the development of domestic climate change regimes.

Comparisons and Interrelationships

The research design enables two levels of comparison: first, comparisons of social actors involved in domestic policy debates on climate change within a given nation-state; and, second, comparisons across nation-states. In other words, this study is designed to look at the relationships between domestic actors within a global environmental system. In the qualitative analysis presented in chapters 4 through 6, I examine to what degree each case fits the expectations of the environmental sociology and environmental state literatures. In addition, I explore the specific dynamics among various social actors at the nation-state level and their interrelations with international actors.

Data Collection and Interview Techniques

While the number and type of interviews varied by nation, those interviewed generally include scientists, government officials, representatives of industry, and nongovernmental organizations. Data collection for each nation took place in two stages. First, from 1998 to 2000, I collected basic background information about each nation's domestic policies on the issue of climate change and the position of each within international negotiations. This information was gathered from secondary sources provided by international organizations such as the United Nations, the Organisation for Economic Co-operation and Development (OECD), the International Energy Agency (IEA), and the Global Environment Facility (GEF); as well as from extant research on the international history of the issue, the political culture of the countries of interest, and media accounts of domestic policy making. Second, from 1999 to 2001, I conducted semistructured face-to-face interviews with key actors involved in the formation of each country's national global climate change regime, focusing on how characteristics of each nation-state interacted to form each country's distinctive regime. In all, I conducted tape-recorded, face-to-face interviews with eighty individuals. Due to the varying complexities of the cases, the numbers were not evenly distributed across countries, being structured instead to understand the relationships across social actors in each nation-state. In addition to those whom I formally interviewed, I gathered further background information by meeting with over sixty-five people involved in international and domestic aspects of the issue of climate change. Table 3.4 lists the number of people formally and informally interviewed in each country.

I began the process of key actor identification by using preexisting studies of the politics of climate change and media reports. During interviews with those identified through these sources, I used a "snowball" approach, asking the interviewees for referrals to other actors involved in the issue. I then contacted the additional individuals, asking them to pass on the names of other appropriate individuals, until I was assured that I had tapped into all of the major relevant positions on climate change in

Table 3.4.   Interviews Conducted in Each Nation

Nation Interviewed	Formal Interviews	Approximate Total
Japan	28	40
Netherlands	23	40
United States	29	50
Total	80	130

each nation. Clearly, due to the size of the community involved in climate change in each nation, I was not able to interview every person working on the issue. Instead, I continued gathering data in each country until the people whom I interviewed no longer provided me with new names of people or organizations that had differing perspectives.

Data for the Japanese case study were collected during two trips to Japan. During my first visit, from June through August 1999, I worked out of the Japanese National Institute for Environmental Studies, in Tsukuba. During my second visit, from February to April 2000, I worked out of the Tokyo Institute of Technology, in Tokyo. In total, I met with over forty people involved in various activities related to climate change in Japan. Of these, twenty-eight were key players within the debate about Japan's approach to mitigating global climate change and were thus formally interviewed. In addition, some follow-up interviews were conducted during the COP-6 and COP6-bis negotiations, in fall 2000 and summer 2001, respectively. Appendix A provides a list of the interviewees and their professional affiliations. The interviews were conducted in Japanese and English. All translations were done by me.[10]

Data for the Dutch case study were collected during two research trips to the Netherlands. During my first visit, from May through July 2000, I worked out of the Department of Environmental Sociology at Wageningen University. During my second visit, from November to December 2000, I worked out of the Department of Public Policy at the University of Amsterdam. In total, I met with more than forty people who were involved in the many aspects of climate change policy in the Netherlands. I formally interviewed twenty-three of these people who were key players involved in determining the Dutch position on climate change. Appendix B provides a list of the interviewees and their professional affiliations. All of the Dutch interviews were conducted in English. Some of the quotations from these interviews have been edited for grammar and clarity.

Data for the United States case study were collected during three research trips to Washington, D.C. The first two trips took place in April and August 2000, during the last year of the Clinton administration. The last trip took place in May 2001, during the first year of the Bush administration. In addition, I met with a number of U.S. representatives at the COP-6 negotiations, in The Hague, and at the COP-6bis negotiations, in

Bonn. In total, I met with over fifty people engaged in the issue of climate change in the United States, formally interviewing twenty-nine of these people who are key players involved in determining the United States position on climate change. I have allotted more space to this case study than to the prior two cases due to (*a*) the difference in size of the political community in the United States versus those in the Netherlands and Japan; (*b*) the complexity of the U.S. position on the Kyoto Protocol; and (*c*) the importance of U.S engagement in the international climate change regime, as this country is responsible for 25 percent of the world's carbon dioxide emissions. Appendix C provides a list of the interviewees and their professional affiliations. Some of the people whom I interviewed agreed to meet under the assumption that they would not be directly attributed. In the cases where I reference those conversations, I cite the person's general affiliation.

Questions asked in the semistructured interviews vary depending on the respondent's position and organizational affiliation, but all interviews address the same range of topics:

- *General background*: the educational and professional background of each person who was interviewed, as well as his or her organization's history when applicable
- *Interpretation of the issue of global climate change:* the perceptions of the interviewee and his or her organization (when applicable) regarding the issue
- *Interpretation of the status of the Kyoto Protocol:* the interviewee's interpretations of the status of the negotiations, as well as this person's interpretation of the status of domestic responses to the issue of climate change
- *Interpretation of the future of the Kyoto Protocol:* the interviewee's opinion regarding the likely outcome of the international negotiations and his or her interpretations of the ratifiability of the protocol once the drafting of the text was finalized.

Analytical Procedures

The semistructured face-to-face interviews were tape-recorded and transcribed, and extensive notes and memos from all of the interviews were kept, thus providing the bulk of the qualitative data set. A qualitative data analysis computer program (NVivo) was used to store, sort, and code transcribed data. In comparing across cases, I used the Boolean-driven method of difference (Ragin 1987), in which patterns among the presence and absence of factors are discerned. I coded each of the interviews several times, and I entered those codes into NVivo. As the patterns across cases emerged, I distinguished between first-order conclusions, those explicitly drawn or stated by the respondent; and second-order conclusions, those I drew from what was said. In so doing, I acknowledge my own role in interpreting the data patterns, as well as subjecting the respondents' claims to additional scrutiny. Unless otherwise noted, all quotes from interviews are referenced as "interview, name year" within the text of the chapters. Results of this analysis are presented in chapters 4 through 6.

SUMMARY AND OUTLINE OF ANALYSIS

Table 3.5 presents a summary of key characteristics of the three case studies. In keeping track of the specifics of each case, the reader may want to refer back to this table when reading through chapters 4 through 6. In these chapters, I analyze the relationship among social actors in each nation-state, as decisions are made regarding national policy and each country's national position regarding the Kyoto Protocol is formed. Chapter 4 describes the Japanese case, chapter 5 describes the Dutch case, and chapter 6 describes the U.S. case. In chapter 7, I integrate the results of these cases and present overall empirical and theoretical conclusions.

Table 3.5. Climate Change Regime Matrix

Country	IV1 State	IV2 Science	IV3 Market	IV4 Civil Society	DV1 Political Outcome	DV2 Material Outcome[a]
Japan	Strong	Central	Collaborative	External-local	Ratify[b]	Increase 7.6% since 1997
Netherlands	Medium	Middle	Autonomous	Internal	Ratify[b]	Increase 9.3% since 1997
United States	Weak	Peripheral	Autonomous	External-national	Nothing	Increase 11.7% since 1997

[a]Measures of emission reductions are calculated from the most recent CO_2 emission inventory (1998) published by the International Energy Agency (2001).
[b]This variable measures political intention and not actual ratification.

NOTES

This chapter builds on Dana R. Fisher and William R. Freudenburg, "Post-industrialization and Environmental Quality: An Empirical Analysis of the Environmental State," *Social Forces* (forthcoming).

1. See also Roberts and Grimes (1997), Roberts (2001).
2. See www.iea.org/about/index.htm for more information (accessed October 1, 2003).
3. See, for example, Daly, Cobb, and Cobb (1989); for a written assessment in the popular media, see Cobb, Halstead, and Rowe (1995).
4. For details on TPES calculations, see www.iea.org/statist/keyworld/keystats.htm (accessed October 1, 2003).
5. Available at www.unep-wcmc.org/protected_areas/data/un_annex.htm (accessed October 1, 2003).
6. See www.yale.edu/envirocenter/research/esi.html (accessed October 1, 2003).
7. See www.ciesin.columbia.edu/indicators/ESI/ESI_01a.pdf (accessed October 1, 2003).
8. Although it is conceivable that the results of this analysis might be biased by the inclusion of the United States, the results remain the same when the U.S. case is removed and the adjusted R-squared decreases to .620.
9. In 1998, the European Union emitted 3,170.5 million tonnes of carbon dioxide, compared to 5,409.8 for the United States (IEA 2001).
10. Grammatical errors were edited from the responses of nonnative English speakers for clarity.

4

State-Led Collaboration in Japan

As a first step in looking more closely at the dynamics of climate change regime formation within countries, we now turn to the case of Japan. To understand the formation of the Japanese climate change regime since the Kyoto Protocol was drafted in 1997, it is important to recognize some of the unique characteristics of the country. Behind all of its domestic and international policy decisions is the Japanese culture itself (for a full review of Japanese culture, see Josai Daigaku 1986; see also, Smith 1983; van Wolfren 1989). Perhaps the most remarkable characteristic of Japanese culture is what the Japanese call *wa*, a notion that is rooted in Confucian and Buddhist thought and emphasizes the "overriding importance of social harmony" within Japanese society (Smith 1983, 49). Even with the continuing evolution of Japanese culture, *wa* continues to be a main tenet.

This inherent theme of social harmony is one of the reasons that the Japanese government has remained under a single party's control for so long. Although Japan has had eight prime ministers in the last ten years, the same political party—the Liberal Democratic Party (LDP)—has maintained a majority in the Diet, the Japanese parliament, almost exclusively since 1955. Even during the one brief period of exception, when the LDP lost its majority in 1993, the political party continued to maintain a high level of power within the Diet by joining in coalitions with other parties. In 2001, for example, the party controlled over 45 percent of both houses of the Diet. The power of the LDP also extends to the prime minister, as the Diet is responsible for selecting the head of the state. In addition, half of the members of the Japanese cabinet, who are appointed by the prime minister, are legally mandated to be members of the Diet. With the LDP maintaining a majority in the Japanese parliament, and thus playing a leading role in determining the prime minister, members of the LDP usually sit in the cabinet. In other words, this one political party continues to wield significant power in Japan.

Another reason for the stability of the Japanese government is the strongly collaborative relationship between the Japanese state and market. In fact, this collaboration contributed to Japan's quick economic recovery after World War II. In the words of Huddle and Reich, "A powerful partnership for central government and private industry . . . concentrated its full efforts on rapid industrial growth" (1975, 26). The partnership was so successful, up until the 1990s, that Japan's economy boomed, growing at rates that Westerners found remarkable (see, e.g., Stone 1969; Forsberg 2000; Grimes 2001).

One of the by-products of this economic growth, however, was the kind of environmental degradation that would have been predicted by the environmental sociology writers whose work was summarized in chapter 1. Perhaps the most vivid consequences of this environmental damage involved the occurrence of pollution diseases throughout the country. In the 1950s and 1960s, thousands of Japanese suffered diseases caused by environmental degradation and pollution. In fact, by the 1970s almost half of the Japanese population thought that they were suffering from a pollution-related disease (for a complete discussion, see Huddle and Reich 1975; McKean 1981).

Although the state and the economic sector were slow to respond, their eventual reactions appear to have been more similar to what would have been predicted by the environmental state theorists. The methods through which citizens voiced their grievances, and the subsequent state and market responses that followed, are consistent with the mechanisms outlined in the theory of reflexive modernization[1] (see, e.g., Beck 1987, 1995, 1997, 1999; Beck, Giddens, and Lash 1994; see also, Offe 1996). In essence, the Japanese government responded to massive citizen demonstrations by passing sweeping regulations to protect the environment in 1970.

Today, evidence increasingly suggests that Japan's leaders no longer see economic growth as being antithetical to environmental protection. As early as 1981, McKean reported that Japan had the strictest pollution regulations in the world, spending more of its gross national product (GNP) on antipollution measures than any other country. This trend has continued; as noted in an article in *The Economist*, for example, Japanese manufacturers embraced the ISO 14,000 standard of environmental good housekeeping[2] "faster than any other country" ("Toxic Waste in Japan" 1998, 60; see also Hideaki 1999). Moreover, in accordance with the expectations of the environmental state literatures discussed in chapter 1, environmental quality in Japan has appeared to improve with economic growth, even with the country's recent economic troubles. In the words of Mitsuda, "the very success of [Japanese] economic growth, the so-called 'economic miracle,' has encouraged the establishment and strengthening of environmentalism in the process of Japanese modernization" (1997, 442).

Despite its general support of environmental issues, the Japanese government was much slower to adopt measures to deal with global climate change. Prior to COP-3, the Japanese did not seem particularly supportive of international climate change regulation that relied on domestic reductions. In 1994, for example, the Japanese government reported to the Organisation for Economic Cooperation and Development (OECD) that it believed that "joint implementation [the opportunity for a country to account for some of its emission reductions inside another] should not be restricted to Annex I parties, but should also be allowed to take place between Annex I and other Parties of the Convention" (International Energy Agency 1994, 107–8). In other words, rather than focus on domestic emission reductions, Japan preferred to focus its early climate change policies in other countries. In fact, throughout the negotiation process since 1994, Japan has maintained its position that carbon sinks and other overseas mechanisms—the framework for which were only finalized after the COP-6bis negotiations—would account for much of its emission reduction commitment.

As late as 1997, Japan remained an active member of an alliance with the United States, Canada, Australia, and New Zealand. This umbrella group, called JUSCANZ, coordinated its negotiating because "these were the developed countries with reasons to want to slow down the European Union. . . . Japan was wholly dependent on imported oil and coal, and already far less energy-intensive than other developed countries" (Leggett 1999, 249). As a result of these factors, Japan, working with the other members of JUSCANZ, pushed for the inclusion of developing country participation in a climate change treaty. Even with their attempts, however, the parties to the Kyoto Protocol finally agreed that the climate change treaty should follow the Framework Convention on Climate Change in its commitments. In other words, these parties ultimately agreed with the Berlin Mandate, which came out of COP-1, in 1995. It had stated that only the "Annex I," or developed, countries would be bound by the first commitment period of the protocol, which was eventually written in 1997 (for a full discussion of the climate change negotiations leading up to 1997, see Gelbspan 1997; Leggett 1999; Paterson 1996; see also, Bodansky 2001).

Although the JUSCANZ umbrella group has continued to maintain a relatively consistent negotiating block throughout the negotiations, there have been cracks in the consensus. One particularly significant difference took place during the COP-2 negotiations, in Geneva, when the United States shifted its position away from the other members of the JUSCANZ block (for a full discussion of the United States' changing position at the climate change negotiations during this period, see Leggett 1999). After the other members of the umbrella group tried to block the change in the U.S. position without success, Japan went along and agreed to the Geneva Declaration, which pushed the parties to work for a legally binding protocol at the next round of negotiations, in Kyoto. As discussed in further detail in chapter 6, this change in the JUSCANZ position was quite significant to the formation of an international climate change regime, and the effects of this decision can still be seen today.

Partially as a result of the specifications of the Berlin Mandate and the United States' leading JUSCANZ members to support the Geneva Declaration, Japan, the host of the COP-3 negotiations, changed its position on climate change. In the words of Taka Hiraishi, a Japanese national who cochairs the National Greenhouse Gas Inventories Task Force Bureau of the Intergovernmental Panel on Climate Change (IPCC), "Japan did not favor a big reduction before Kyoto. Now, Japan believes that there is a need for a big reduction" (interview, 1999). Some people say that the change in Japan's position was also a response to the European Union's pressure regarding emission reductions. Going into the COP-3 negotiations, the EU supported the Kyoto Protocol, requiring a 15 percent reduction in greenhouse gas emissions below 1990 levels. In the words of Yasuyoshi Tanaka, the environment writer for the third largest newspaper in Japan,[3] the *Mainichi Shimbun*: "If it were not for the EU decision, the Japanese would not have agreed. International pressure played a role. The Japanese are weak to international pressure. . . . Because the EU was pushing for more [emission reductions], Japan changed its position" (interview, 2000).

International pressure has played a role in many decisions regarding Japan's position on climate change. Consistent with the global environmental system, this interrelation between the domestic and international sides of the process of climate change regime formation has significantly contributed to political outcomes both inside Japan and in the larger global context. In addition to the change in the JUSCANZ position and the international pressure from the EU, Japan was particularly committed to the outcome of the Kyoto Protocol, as it was the host country for the very conference that officially drafted the protocol; the Kyoto Protocol itself even holds the name of one of the most sacred cities in Japan.

Although some international environmental groups have accused Japan of being one of the countries that is trying to do everything it can to wreck the treaty, Japan's commitment to the mitigation of climate change appears to be genuine. Since the official drafting of the Kyoto Protocol, in December 1997, Japan has been actively promoting the regulation of climate change, and the Japanese government has passed legislation in response to the protocol. The first step in the establishment of a climate change regime in Japan was the "Law Concerning the Promotion of the Measures to Cope with Global Warming" (law number 117), which was passed in 1998. The law "aims to promote the measures to cope with global warming through e.g. defining the responsibilities of the central government, local governments, businesses and citizens to take measures to cope with global warming, and establishing a basic policy on measures to cope with global warming, and thereby contribute to ensuring healthful and cultural lives of present and future generations of people, and to contribute to the welfare of all human beings" (Environment Agency of Japan 1998, 1). Beyond this preliminary legislation, the Japanese government has debated a number of specific measures to achieve its 6 percent reductions. As mentioned, Japan has planned to account for 5.5 percent of its reductions below the 1990 level through growing forests—otherwise known as carbon sinks—and other actions outside its borders, using what are called the "Kyoto Mechanisms." This lack of significant domestic emission reductions in the Japanese climate change plan has led activists from around the world to criticize the Japanese position. Commenting on this scenario, and others throughout the chapter, is Shuzo Nishioka, the executive director of the Japanese National Institute for Environmental Studies; a professor of media and governance at Keio University; and the project leader of the climate change group at the Institute for Global Environmental Strategies (IGES), one of the leading Japanese think tanks working on climate change. Nishioka points out that, once the protocol is ratified in Japan, domestic actions will be taken.

In addition, even though Japan intends to meet the 6 percent reduction below its 1990 levels predominantly through nondomestic measures, it has been one of the more successful countries in stabilizing domestic emissions since 1997. Emission levels in Japan have supported Leggett's statement (1999) that Japan is one of the most efficient of the developed countries. Even before the Kyoto Protocol was drafted, per capita CO_2 emissions in Japan were less than half as high as in the top emitter in the world, the United States ("Global Warming Update" 1996). According to World Resources Institute calculations (1994, 208), Japan's total greenhouse gas emissions in 1990 were only a quarter as high as those of the United States. Even with its high levels of efficiency, however, emissions in Japan have still grown since 1990. As of 1998, Japan had increased its overall CO_2 emissions to 7.6 percent above 1990 levels (IEA 2000b). Table 4.1 provides both overall and per capita comparative emissions data for

Table 4.1.   Overall Emissions and per Capita Emissions of Carbon Dioxide for
Selected Nations (in million tons of CO_2)

	Japan	Netherlands	United States
Overall emissions, 1990	1,048.5	156.8	4,843.8
Overall emissions, 1998	1,128.3	171.4	5,409.8
Change (%)	_7.6_	_9.3_	_11.7_
Per capita emissions, 1990	8.4	10.2	18.6
Per capita emissions, 1998	9.0	11.1	20.8
Change (sum)	_0.6_	_0.9_	_2.2_

Source: International Energy Agency (2000b)

Japan, the United States, and the Netherlands, since 1990. As can be seen, although Japan's emissions have increased since 1990, it has the lowest level of growth of the three countries.[4] Perhaps of more interest, however, is the data on the per capita level. Since 1990, Japan's per capita CO_2 emissions have increased only by 0.6 million tons carbon dioxide per capita.[5] As discussed later, at least part of these emission reductions are attributable to the severe economic recession that hit Japan in the early 1990s.

This chapter follows Japan's responses to the Kyoto Protocol leading up to the COP-6bis, in July 2001 in Bonn, Germany. It begins with the signing of the protocol in 1997, looking at the various social actors involved. Consistent with the Japanese notion of _wa_, the Japanese case presents an example of a strong state working closely with scientists and a collaborative market sector to bring about positive policy outcomes. This chapter concludes with a summary of Japan's position on the Kyoto Protocol at the time of the COP-6bis negotiations, in July 2001.

RESULTS: POST-KYOTO STATUS

Science

In contrast to the United States, where the issue of the validity of the science has long been at the center of the climate change debate, the science of climate change has been basically accepted in Japan. Of all of the people whom I interviewed in Japan, not one person challenged the validity of the science of climate change. Instead, there was just one skeptical response to the issue, by Kimiko Hirata, a leader of the Kiko ("Climate") Network, a nongovernmental organization (NGO) that coordinates environmental groups working on climate change; however, the organization scarcely endorses the skepticism: "Most of the people believe that climate change is happening. Of course there are the skeptics on this issue, but they do not have an impact" (interview, Hirata 1999). In general, as can be summed up in the words of Nishioka, "There is no real opposition to the 'scientific view' here in Japan. There is no suspicion about science. It is accepted" (interview, 1999).

Hiraishi, however, presented a viewpoint that was only slightly different: "The focus is on the political and negotiation even though more discussion of the scientific argument is needed" (interview, Hiraishi 1999). Although scientists in Japan acknowledge that there are uncertainties regarding the science of climate change, they tend to agree that, in the words of Tsuneyuki Morita—the head of the social and environmental systems division at the National Institute of Environmental Studies, a coordinating lead author for the Intergovernmental Panel on Climate Change (IPCC) *Third Assessment Report for Working Group III*, and an economics professor at the Graduate School of Decision Science and Technology at the Tokyo Institute of Technology—"At this moment, scientific knowledge is at a very, very low level about the climate. We need a lot more time but we cannot wait. . . . Policy should be promoted simultaneously" (interview, 1999). In other words, Morita supports Japan's following through with the precautionary principle (outlined in chapter 2).

To some degree, due to the role that academics play in policy making in Japan, the science and the politics of the issue have become conflated. Many of the scientists whom I interviewed hold positions as academics and within of political agencies. One such example can be seen in the case of Yoichi Kaya, a professor of the graduate school at Keio University. In addition to holding an academic position, Kaya is the chair of two energy-related committees for the Ministry of International Trade and Industry (MITI); he is also the vice chair of the integrated council of the government that coordinates policy strategy development on global environmental issues. In his own words, "I have been connected quite closely to the government, but still, I am a scientist" (interview, 1999). Similarly, Morita sees himself as an "interface between policy making and science . . . a go-between who interprets the line between science and policy" (interview, 1999). In other words, Morita identifies himself as bridging the science–policy continuum. Simply put, these scientists are not unique in their affiliations. "Many members of the Central Environment Council and the joint panel are at the same time members of MITI's advisory councils" (Ouchi 1998, 26).

In sum, science plays a central role in policy making in Japan—so much so that scientists actually hold positions on government councils. The scientists working within Japanese academia are the same people who are leading the national labs to conduct research on climate change, and in many cases, they are the leaders of the governmental committees that are deciding national climate change policies. These close relationships between science and policy have significant implications on the efficacy of environmental policy in Japan; at a minimum, because leading scientists are linked with the government, scientific findings are easily diffused to policymakers.[6]

The State

This connection between science and policy needs to be kept in mind as we turn to the role that the Japanese state has played in dealing with the issue of climate change. Not only are the people involved in Japanese policy making often the same as those working on the science of climate change, but actors within the state in Japan see the relationship between the issue's science and policy as inherently related. In the words of Kazuo Matsushita, an ex-member of the Environment Ministry[7] who now serves as the acting vice president of the government-funded Japanese think tank the Institute for Global Environmental Strategies (IGES), the goal of this type of work is to "translate these global environmental issues into domestic policies" (interview, 1999).

In general, the Japanese government is perceived as being very strong and thus responsible for taking the lead in the issue of climate change. As Harumi Suda, a leader of the consumer and environmental movement in Japan and the director of the Shimin Undo ("National Citizen's Movement") Center states, "Because the government is strong, if the government does not lead, lifestyles cannot change" (interview, 1999). Although the basic expectation in Japan is that the government is responsible for dealing with environmental issues such as climate change, Morita points out that "the government cannot do much about what people think, and they are very critical" (interview, 1999).

Perhaps partly because of the belief that citizen lifestyles may not easily change as a result of governmental policy, climate change mitigation in Japan is based on a top-down approach that relies heavily on government action. In the words of Akiko Domoto, a former member of the Japanese Diet who became the mayor of Chiba prefecture in 2001, "Implementation of numerical targets is a priority" (interview, 1999). This opinion is echoed in similar statements by Japanese bureaucrats, such as Hironori Hamanaka, the director general of the Global Environment Department of the Environment Ministry and one of the heads of the Japanese climate change negotiating team: "The goal is, of course, to implement our own policy effectively . . . and to reduce our emissions of greenhouse gases" (interview, 1999).

Although the Environment Ministry is responsible for dealing with and regulating environmental issues in Japan, MITI is responsible for energy policy. As such, MITI is responsible for instituting the measures that deal with global climate change. This distinction between the regulating bodies in Japan leads some outsiders, such as Suda, to claim that "because the different governmental agencies are fighting against each other, it is hard to get things accomplished" (interview, 1999). Still, even though some level of infighting between the agencies may very well be taking place, the notion of harmony and *wa* generally prevails. For example, Morita insists that "they both agree that something should be done" about global climate change (interview, Morita 1999).

On June 19, 1998, the Global Warming Prevention Headquarters of the Japanese government, which includes cabinet members and the Japanese prime minister himself, published the *Guideline of Measures to Prevent Global Warming: Measures towards 2010 to Prevent Global Warming.* The guideline reports the measures that the government plans to take to achieve the 6 percent emission reductions stipulated in the Kyoto Protocol. In essence, the government plans to "strengthen efforts which can attract the participation and cooperation of every social actor, mobilize every possible policy measure, and promote comprehensive actions in a systematic way in order to steadfastly achieve the reductions" (Global Warming Prevention Headquarters 1998, 1). The report continues by pointing out the government's main goals of achieving a 2.5 percent reduction in domestic emissions by "promoting measures relating to both energy supply and demand focusing on promoting energy saving, introduction of new energy and the construction of nuclear power plants" (2). As a means of providing a cleaner energy source to support Japanese economic growth, the Japanese government's global climate change reduction plan included "an energy supply forecast with 20 additional nuclear reactors as its centerpiece" (Ouchi 1998, 25).

When dealing with energy policy, one of MITI's main concerns is energy security. As an International Energy Agency report points out, "Japan's limited indigenous resources translate into high dependence on energy imports. . . . Security of energy supply is a policy priority, since energy demand is expected to rise as a stimulus to economic growth" (1994, 110). This priority relates directly to the fact that the central aspect of the Japanese government's response is the construction of new nuclear power plants. Given the lack of any indigenous energy options in Japan, the government has chosen to invest in a domestic energy option that is considered clean in terms of greenhouse gas emissions. Hiraishi justifies the Japanese government's decision: "Is the 6 percent reduction feasible? Without nuclear power it is not possible" (interview, Hiraishi 1999).

Despite the notion of *wa*, however, this plan to increase the number of nuclear plants in Japan has been criticized by many sectors of society. Particularly in the aftermath of the nuclear accident that took place at the Tokaimura power plant in September 1999, many Japanese have been less

supportive of nuclear energy in Japan.[8] In addition, the decision to build
these nuclear plants was based on models that predicted energy consumption extrapolated from the Japanese economy of the early 1990s. As
a result of the decreased economic growth in Japan—or, as many Japanese would say, since the "bubble burst"—energy consumption has
slightly decreased.[9] With the stabilization of energy consumption, advisors to MITI, such as Kaya, have pointed out that "the environmental target [of the Kyoto Protocol] will be realized . . . because of the reduction in
economic growth" (interview, 1999). In 2000, the largest English-language
newspaper in Japan, the *Japan Times*, reported emission reductions that
substantiate Kaya's prediction. "Environment Ministry officials say the
decline in emissions is due largely to a drop in the industrial sector, at-
tributing the reduction in part to economic stagnation" ("Greenhouse Gas
Output Declines" 2000, A6)

Concurrent with new energy policies being coordinated by MITI, the
Environment Ministry designed and implemented the Law Concerning
the Promotion of the Measures to Cope with Global Warming. This law
involves a nationwide inventory of greenhouse gas emissions, conducted
on the local level. This plan also, in the words of Hamanaka, "requires national government and all local authorities, local government and also
3,200 cities, towns and villages . . . to draw up their own plans . . . to reduce greenhouse gases" (interview, 1999). In addition to the inventories
and plans, the law creates a national center for climate change that will
work with nongovernmental organizations (NGOs) and citizens.

The law has been criticized by members of NGOs—such as Naoyuki
Hata of the Citizen's Forum, who states that it is "weak" (interview,
1999)—because these steps are only preliminary and do not go far enough
to meet Japan's emission reduction targets set in the Kyoto Protocol. Others, such as Yasuko Kameyama,[10] a climate change researcher at the National Institute of Environmental Studies, see it as "the starting point . . .
[to help us know] from where emissions come" (interview, 1999). Furthermore, the Japanese government has been debating other measures. In
an interview at the climate change negotiations in The Hague, in November 2000, Hamanaka outlined a number of climate change mitigation measures that are being considered by the government for implementation after the protocol is finalized and ratified by the Japanese government.
These measures include the possible amendment of the energy conservation law to include emissions trading, as well as the possible introduction
of a "green tax" that would apply to consumers and small businesses. The
Institute for Global Environmental Strategies, in fact, held a forum in November 2000 to review Japanese climate change mitigation policy options,
including those mentioned by Hamanaka (IGES 2000). However,
Hamanaka also stated, in an interview in November 2000, that in the end

the implementation of some or all of these measures is "based on the out-come of the [COP-6] conference," which would not be known until after the COP-6bis meeting, in July 2001.

Until then, the Japanese government was maintaining its limited en-ergy policy and local projects that addressed the issue of climate change until the outcome of the negotiations were known. Japanese and interna-tional NGOs continue to criticize the Japanese position for being too timid, and the Japanese government for "just waiting until after ratifica-tion" (interview, Hirata 2000). Nevertheless, the Japanese government has taken preliminary steps to consider how it will meet its commitment and has actually reduced emissions since 1997, while other countries have not.

Industry

Given Japan's strong state, it is not a surprise that Japanese society is also well known for its close ties between the state and the industrial sectors (for a full discussion, see van Wolfren 1989). The state tends to work closely with industry, which in turn collaborates to determine how to im-plement necessary regulations. A report about the Japanese government's Central Environment Council, which advises the government on issues such as global climate change, says that "the reality on the council level is one of close cooperation between the Environment Ministry, MITI, and business" (Ouchi 1998, 26). The regulation of climate change in Japan is a case in point in that the government has only taken steps that are ap-proved by industry. Hirata stresses this point, saying that "there is always the government choosing industry" (interview, 1999). In other words, NGOs are critical of the fact that the state is sensitive to Japanese indus-try and that, when given a choice between industry's desires and the en-vironment, the environment tends to lose.

Although the Japanese government has committed itself to implemen-tation of the Kyoto Protocol and has even passed a preliminary law as the first step in regulating climate change, it has avoided imposing regula-tions on industry. Hamanaka describes the process of environmental law-making in Japan and the present climate change law:

> In our country, the industries themselves accept their own action plans with their goals. . . . We thought it would be premature to impose legal require-ments for emission reductions, so basically this law just promotes voluntary actions to establish plans for reduction of greenhouse gases. . . . This is just a voluntary action . . . [for] industry. They are not legally required to achieve the goal. (interview, 1999)

As this high-ranking governmental official points out, the Japanese gov-ernment is trying to provide industry with a guide to its priorities. The

Japanese government considers voluntary commitments from industry "major policy measures to be applied" (Akio Morishima, head of the Central Environment Council, as quoted in Corliss 2000, A9). In other words, the government uses voluntary laws, such as the Law Concerning the Promotion of the Measures to Cope with Global Warming, to warn industry that it needs to come up with a plan to deal with environmental issues such as climate change.

Still, unlike what might be expected by critical or skeptical analysts who see these types of voluntary measures as merely symbolic actions by the state (see, e.g., Edelman 1964), the Japanese government's warning seems to be taken to heart by Japanese industries. Nishioka states that industry focuses on these "top-down approaches" (interview, 1999). At the same time, both government and industry are interested in reducing energy consumption for an altogether different reason: Japan has a very small indigenous supply of fuel.

All in all, the strategy of the government's warning industries about its priorities seems to have been reasonably effective in promoting voluntary agreements in industries. Nippon Steel, for example, voluntarily committed to decreasing its energy consumption. In the words of Teruo Okazaki, the senior manager of the global environmental affairs department of Nippon Steel Corporation, "the energy savings in our production line is one of the most important things for us. . . . Our target is 10 percent for energy savings so from 1992 to 2010, 10 percent of our total amount of energy consumed will be saved" (interview, 1999). Similarly, Hajime Ohta, the executive counselor of Environment, Energy and International Economic Affairs of the Keidanren ("Japanese Federation of Economic Organizations"), which is responsible for 60 to 80 percent of the country's gross national product, says that the "business sector is responsible for 40 percent of Japan's carbon dioxide emissions. They will do what they promised" (interview, 1999). Even though these are voluntary measures, many actors in industry do not feel that they have a choice. Ohta goes on to say that the Keidanren recognizes that if its member companies "did not do it on their own, the government would regulate them" (interview, 1999). In a related editorial, Ohta points out that through the increased pressures for voluntary agreements, "the government is effectively tightening its control over corporate activities" (Ohta 1998).

Even with these pledges, some NGOs have criticized the voluntary measures. In the words of Yurika Ayukawa, the climate change campaign officer for the World Wide Fund for Nature (WWF), Japan, "There is no place for a third-party evaluation or verification team. . . . They will say they will do this and I think they will do it, but who knows?" (interview, 1999). Even while she complains about the evaluation of these agreements, there is a sense that these agreements are effective and that industry will achieve its goals. The director general of the Japanese Environment Ministry is even more certain of industry's commitments to climate change: "They are, all of them, well-known companies, and they already made their plans public. I think that the CEOs are under heavy pressure if they fail to achieve that goal. . . . I think the likelihood or the possibility for them [the companies] to reach their own target would be . . . large" (interview, Hamanaka 1999). This behavior is consistent with other aspects of Japanese culture in that harmony (or *wa*) and saving face are central components of social interaction in Japan (e.g., van Wolfren 1989; Smith 1983). Positive environmental outcomes of these agreements can already be seen in the Keidanren report that shows a 9.5 percent decrease in the emissions of Japan Iron and Steel Federation and the 9.7 percent decrease in the emissions of the Cement Association of Japan between the years 1990 and 1998.[11]

In response to such tangible evidence of receptiveness, there is an overall approval of the role that industry is playing in dealing with climate change in Japan. In particular, Japanese industry has been lauded as being exceptionally efficient. Some have even called Japan "the most energy efficient country of the OECD" (interview, Ohta 1999; see also Leggett 1999). The big criticism of Japanese companies comes not from their efficiency standards and ability to implement commitments to such standards but from their ever-increasing production and the growing pressure for consumer consumption. Again, NGOs criticize Japanese industry. For example, Ayukawa provides a very different criticism here, stating that companies "will reduce their energy consumption. They will make their products more energy-efficient, but they will make more. That is what they are saying. And that is what the projection is: we will be using more computers, more phones . . . and we will be using more energy in 2010" (interview, 1999). Even with this criticism, Japanese industry, by initiating its own voluntary measures, is working with the government to respond proactively to the issue of climate change using a top-down approach.

At the first part of the COP-6 climate change negotiations, in fall 2000, and prior to the ratification of the Kyoto Protocol in Japan, Hamanaka discussed industry's role in emission reductions. He pointed out that industries and the Keidanren, as the largest association of industries in Japan, are cautious about taking any further steps. "They seem reluctant to move forward and sign an agreement with government. They think that other things should be done first on a consumer level" (interview, Hamanaka 2000). Given that the Japanese economy has been suffering for many years and that the protocol's entry into legal force has been uncertain, it makes sense not to implement costly measures. However, with Japan's strong state and its close ties with industry, implementation of measures to achieve the country's emission reductions by 2008 to 2012 seems feasible.

Civil Society

In contrast to industry, which has been aggressively working to deal with the issue of global climate change in Japan, civil society has paid much less attention to the issue and has been relatively external to the domestic climate change discussion. As mentioned, the industrial sector in Japan has an economic incentive to follow the issue of climate change and respond to the issue to avoid governmental regulation. For citizens, the situation is different. Makoto Ikeuchi, the managing director of the Hokkaido Seikatsu ("Lifestyle") Club, one of the largest consumer cooperatives in Japan, points out that "because the economy is bad, many people are not thinking about the environment. People are thinking about prices and money" (interview, 2000). When it comes to climate change, Japanese citizens do not appear to be particularly motivated to put pressure on the institutional structures of the state or the market to bring about environmental changes as they did in the 1970s to stop the pollution diseases.

In addition to the financial problems facing many Japanese citizens during the recession that began in the 1990s, the complexity of the issue of climate change also contributes to the lack of citizen activity. Surrounding the COP-3 negotiations in Kyoto, the Japanese media were deeply involved in covering the issue of climate change; the major newspapers in Japan covered the issue of climate change on an almost daily level (interview, Tanaka 2000). Although there was a good deal of media coverage of the issue in 1997, coverage has since dropped significantly. Given the conclusions of scholars such as Anderson (1997), who find that media coverage of environmental problems does not tell people what to think but rather what they should think about, public consciousness about the issue of climate change is presently quite low in Japan. In the words of Ayukawa, "People do not know about global warming so much any more" (interview, 1999).

Beyond the lack of media coverage and financial troubles, many of the people whom I interviewed pointed out that the complexity of the issue of climate change has limited the citizenry's understanding of the issue. Environmental experts such as Kazuo Matsushita—an ex-member of the Environment Ministry who now serves as the acting vice president of one of the leading Japanese think tanks working on climate change, the government-funded IGES—states that "for the general public, I suppose it would be very difficult to understand the overall significance and importance of the climate change issue" (interview, 1999). Similarly, Hirata points out that "global climate change is very tough to understand, only about 10 percent of the people do" (interview, 2000). Writers for the newspapers themselves echo such opinions. In the words of Tanaka, "Japanese citizens cannot get all of the issues involved in global climate change" from reading the newspapers, since understanding the issue requires more depth than the amount of space available in a newspaper can provide (interview, 2000). Although citizens are reported to have a low level of knowledge about global climate change, national policy making continues to move forward due to the strong relationship between the state, industry, and science.

Still another reason for the lack of citizen participation in the issue of climate change may be the structure of Japanese civil society itself. Many of the people whom I interviewed, including citizen activists, pointed out the weakness or lack of an active citizenry in Japan. In the words of Suda, "Citizens' involvement is not Japanese society's style" (interview, 1999). The weakness of the citizenry leads some, such as Ohta, to say that the "public always tries to find somebody to take care of them" (interview, 1999). Most people in Japan agree that the citizen activism against the pollution diseases of the 1970s no longer exists. "In Japan, [citizens] receive information from the government but not from the bottom-up. There are no protests now" (interview, Hirata 2000).

Although the public is not very engaged in the issue of climate change in Japan, a number of nongovernmental organizations (NGOs) are working on the issue of climate change. Mie Asaoka, the president and founder of Kiko Network, points out that some of these organizations work specifically to "widen the public's consciousness" about the issue (interview, 1999). Other NGOs negotiate with the national government about Japan's international climate change position, provide input into national legislation regarding climate change, and work with industry. One of the NGOs most active in working with industry is the World Wide Fund for Nature (WWF), Japan. Ayukawa describes the WWF's work with industry: "Our future is to be working with industry, so we have this industry approach, and I am trying to gather a group of companies who are progressive or very interested and dedicated to the prevention of global warming" (interview, 1999). Even with this involvement in national issues and industry programs, many of the people whom I interviewed identified Japanese civil society as being relatively external to national politics in Japan. In her comments on the citizenry in Japan, Asaoka stated that "the citizenry is so weak. . . . It is not so much that the citizens' movement is very weak, but it is different" (interview, 1999).

Perhaps the most appropriate interpretation of the uniqueness of civil society in Japan is that it has an obvious spatial component. Members of the Seikatsu Club best describe it: "There is no *shimin undo* [national citizen's movement] in Japan, but there is a *jumin undo* [local people's movement]. Local movements exist, but there is not much on the national level" (interview, Demura 2000). Mie Asaoka of the Kiko Network builds on this idea: "Citizens are only active at the local level and they only have small involvement on the national level. The local level is where it can happen.

. . . NGO action is within a small community. They want it to be larger, but there is not much on a national level" (interview, Asaoka 1999). These statements by NGO leaders in Japan are consistent with some of the findings of Mitsuda in his research on environmentalism in Japan (1997; see also Mitsuda and Fisher 2000).

When discussing the specific character of civil society in Japan, many of the people whom I interviewed discussed the role that locally based lifestyle changes could take in mitigating climate change. One of the most well-known movement leaders in Japan, Harumi Suda, states that "the best way to change the system is through the community. Changing citizens' lifestyles could happen through community culture" (interview, 1999). In short, the results of my interviews with representatives of national environmental NGOs suggest the need to look at more locally based organizations involved in global climate change.

To gain a better understanding of the dynamics of civil society in Japan and its action with regard to climate change, I decided to follow the recommendations of a number of NGOs in Tokyo that mentioned a local project in Hokkaido, the northern island of Japan. The project was described by WWF, Japan, as "one big consumer group . . . saving electricity in their own group and . . . supporting a wind turbine in some other area" (interview, Ayukawa 1999). Begun as the response to a proposed nuclear power plant in Hokkaido, the Seikatsu Club cooperative created a project to decrease energy consumption and develop an alternative source of energy. The Seikatsu Club is now one of the largest cooperatives in Japan, with over 250,000 family memberships throughout the country. As the alternative electricity project gained support, it divested from the Seikatsu Club and is now called the Green Fund. Nonetheless, the fund continues to be run by members of the cooperative. "The fund charges its members for their electric bills plus an additional 5 percent of their bill, which is put in the 'Green Fund.' The fund then pays the electric companies for its members' electric bills. Any resident of Hokkaido can join this fund by paying an annual membership fee."[12] Once enough money is collected, it will be matched by a governmental subsidy to fund the construction of one windmill. Electricity generated from the six-megawatt windmill will go to the electricity company that distributes electricity in Hokkaido.

When I met with the members of the Seikatsu Club and the Green Fund to learn about their wind-energy project, I asked how the goals of their project are different from the various wind-energy parks being developed by the government on Hokkaido that are supported by MITI and jointly funded by the government (33 percent) and industry (66 percent). Most of the organization's members had little awareness about the government's plans; in fact, most of the members were visibly surprised by the growth of wind energy on Hokkaido. In response to a question about the significance of their project, Ryuriko Demura, a member of the Seikatsu Club board of trustees and an organizer of the Green Fund, responded, "The Green Fund project is different from the government's because of where the money comes from" (interview, 2000). It is true that the government has windmill projects in which, in the words of a Hokkaido "New Energy" representative, "no citizen money is involved" (interview, Mizushima 2000). The Green Fund's project, in contrast, is motivated, developed, and funded by citizens. Instead of working with the government to promote wind energy in Hokkaido, these citizens have chosen to pursue their own separate program.

As of March 2001, enough money had been collected by the Green Fund to begin construction on the planned windmill. It is unlikely, however, that the windmill will become a major source of electricity for the citizens who supported it—its six megawatts will represent less than .00034 percent of Japan's 1998 electric demand (IEA 2000c).[13] Still, the project does provide a positive example of an active, locally based civil society in Japan. Like many social movement organizations that are constrained by resources, the organizations involved in the windmill are focusing their attention on a project that is firmly external to the social complexes of the Japanese state and industry. It is unclear, however, if local citizen groups might have a larger effect on global climate change were they to work with, or between, the government and industry. Although changing consumer demand is only one method of reducing emissions—and a relatively slow one at that—lifestyle changes such as the ones begun by members of the Green Fund in Hokkaido could contribute to the domestic climate change regime in Japan.

CONCLUSION

The case of the Green Fund project in Hokkaido provides data for one of the main findings of this case study of Japan: the clear spatial component to civil society in Japan. Although there is not a strong Japanese civil society working on the national level to deal with a global issue such as climate change, there is a locally based civil society in Japan that is engaged in the issue of climate change external to other Japanese social actors. With this relatively minor exception, my interviews yield a striking consensus regarding the absence of citizens in the national debate and in playing a role in the mitigation of global climate change in Japan.

This conclusion is consistent with earlier research on Japan, reporting a "failure to achieve a civil society . . . related to the particularism of earlier village society" (Knight 1996, 239). In other words, the absence of a citizen presence in the national work on climate change and the strength of local

jumin-level movements is the result of the history of Japanese society itself. It was only after the disfiguring, pollution-related diseases struck
people throughout the country that Japanese citizens demonstrated
against industrial pollution. Since that time, the Japanese government has
maintained a strong leadership position over many environmental issues,
including global climate change, while collaborating with industry and
keeping science central to the decision-making processes. In the words of
Tanaka, "The government must communicate the point that climate
change is very important. Industry must improve its materials and efficiency . . . and citizens must choose efficient products even if the products
are more expensive" (interview, 2000).

In many ways, the case of Japan provides a clear example of a strong
state that is able to collaborate. That the Japanese state is also collaborative provides a case of what Evans (1995) calls "embedded autonomy." In
Japan, social actors have distinctive roles in the climate change regime.
The citizens' role, however, is generally external to the policy-making
process and predominantly local.

Although the history of Japanese government responses to environmental concerns is consistent with the theory of reflexive modernization,
the present global climate change regime in Japan—with its strong state,
collaborative market, and central role for science—comes closer to fitting
the model of ecological modernization. Specifically, the Japanese climate
change regime illustrates a top-down approach to dealing with the issue.
In fact, some scholars have suggested that the theory may best describe
Japan itself (e.g., Dryzek 1997; Giddens 1998).

Still, even though Japan's climate change regime may come close to fitting the outline of ecological modernization, there is one significant exception: proponents of ecological modernization theory expect social
movement organizations to work with the state and industry to build new
coalitions (e.g., Mol 2000a). To date, these new coalitions are not forming
in Japan. Thus, the preliminary results of my Japanese research find actors
from science, industry, and the state leading the way in making environmental protection possible without Japanese civil society playing an internal role in the policy and decision-making processes.

In other words, these data from Japan suggest flaws in the argument
about how political possibility is created. This absence of an active civil
society in Japan, at least at the national level, obviously challenges the theories that say there must be significant social movements and a strong
civil society to have ecological modernization. One possible interpretation
of the Japanese case is that the magnitude of past environmental mistakes
created a political consensus around avoiding them in the future—a consensus that evidently extends even to the industries that are in a position
to create or avoid such mistakes. Perhaps a more fitting interpretation is

that the particular configurations of social actors in Japan, like the international pressure mentioned earlier in this chapter, support the notion that the political outcomes surrounding the issue of climate change are dependent on the global environmental system.

At the first part of the COP-6 negotiations, advisors to the Japanese government such as Nishioka stated that the emission reduction target agreed upon by Japan in 1997 was a "huge mistake" (interview, 2000). He argued that Japan would not be able to meet its commitments without the protocol's inclusion of flexible mechanisms, such as carbon sinks. At the same time, however, Nishioka said that Japan was dedicated to the Kyoto Protocol. Even if there were a change in the U.S. position—which actually came to pass in March 2001—Nishioka argued that it "does not matter" to Japan's intention to ratify the treaty.

Two months after the U.S. position changed, the newly elected Japanese prime minister contradicted Nishioka's claim. During meetings with leaders in the United States and the European Union, and prior to the COP-6bis negotiations, Koizumi indicated that "Japan's ratification of the pact is linked to a change of heart by Washington" ("EU Won't Alter View on Kyoto" 2001). Given Japan's economic problems and the stated resolve of the new prime minister to solve those problems, it was unclear if Japan would continue to build a climate change regime in the face of recent developments. It was also unclear whether there would be changes in the roles of social actors involved in determining how Japan could meet its emission targets. All that was certain was that Japan had begun to develop an effective climate change regime during an economic recession and that final decisions regarding the future of the Kyoto Protocol would take place at the end of the COP-6 negotiations.

NOTES

An earlier version of this chapter was published as "Beyond Kyoto," in *Global Warming and East Asia: The Domestic and International Politics of Climate Change,* ed. Paul G. Harris (New York: Routledge Press, 2003): 187–206. This research was made possible through the support of the U.S. National Science Foundation.

1. For the sake of simplicity, I am including Beck's work on the risk society (1999) and subpolitics (1997) within the literature on reflexive modernization. Although Beck has used differing terms to describe these theories, they are consistent in their main points.

2. The ISO 14,000 are the environmental standards of the international organization for standardization, which promote "the development and implementation of voluntary international standards, both for particular products and for environmental management issues" (www.epa.gov/owm/iso14001/isofaq.htm; accessed January 7, 2004).

3. Based on circulation rates for 1999, posted on www.business.vu.edu.au/bho2250/Top20Media/TopmediaAsia.htm#Japan (accessed October 1, 2003).

4. Even though the European Union has just about stabilized its emissions and was only .9 percent above 1990 levels in 1998, the EU has achieved this goal through a unionwide burden-sharing agreement that includes technology transfers to the former East Germany.

5. Although comparable data are only available through 1998, Japanese emissions are expected to have decreased even further since then.

6. At the same time, the actual role that scientists play in policy making can be problematic. One such example can be seen in the case of the pollution diseases, where scientists recognized the linkage between the pollution and the disease but their findings were covered up by the government.

7. Although the Environment Agency only became a ministry of the Japanese government in 2001, it is referred to as the Environment Ministry throughout this chapter, for simplicity.

8. It is important to note that, even before the Tokaimura accident, Japanese citizens were concerned about nuclear energy.

9. The United States Energy Information Administration reports that Japanese primary consumption totaled 21.75 quadrillion BTU in 1997 and 21.48 quadrillion BTU in 1998. In other words, there was a decline of 1.2 percent (EIA 1999, 178).

10. In fall 2001, Yasuko Kawashima changed her last name to Kameyama. For the sake of consistency, she is referred to as Kameyama throughout the chapter.

11. See www.keidanren.or.jp/english/policy/pol114/attachment1.html (accessed October 1, 2003).

12. See www.cnic.or.jp/english/topics/energy/renewable/%5B2%5Dwind-hydro.html (accessed October 1, 2003).

13. This calculation is based on 6,000 hours of operation per year.

5

Market Innovation with Consumer Demand in the Netherlands

Similar to the notion of *wa* in Japan, Dutch politics are known for their sense of collaboration. In turning to the Dutch case, it is important to note the easygoing character of the people and institutions within this postcolonial state. Acceptance and collaboration are central to the Dutch way of thinking, as can be seen by the Netherlands' rank as the first country to legalize euthanasia and as one of the only countries to legalize same-sex marriage. Even tourist guidebooks remark on the extent to which such flexible and consensual approaches characterize the practice of politics in the Netherlands (see, e.g., Colijn 1984; Catling 1995).

Beyond the more popular sources, academics have also written about these characteristics of the Dutch people. Perhaps Lijphart, in his seminal work on the "politics of accommodation" in the Netherlands, best describes Dutch politics:

> Dutch politics is characterized by "mutually reinforcing," "superimposed," "congruent," and "parallel" rather than "crosscutting" affiliations and organizational patterns; class and religious cleavages separate self-contained "inclusive" groups with sharply defined "political subcultures"; and there is a multiparty system with considerable "interpenetration" within each sphere among parties, interest groups, and the communication media. But Dutch democracy is eminently stable and effective! (Lijphart 1975, 15)

Although Lijphart identified a level of breakdown in this type of accommodation in the late 1960s, other Dutch scholars have disagreed. In their work on environmental politics in the Netherlands, for example, van Tatenhove, Arts, and Leroy speak directly to the continuation of this type of politics: "As a result of the consensus politics in the Netherlands, there is less necessity to develop an oppositional strategy in the Netherlands" (2000, 5).

In contrast to Japan's variable position on global climate change, the Dutch government has maintained the same position throughout the negotiation process. Consistent with its political position, the country has adopted many progressive domestic policies to deal with global warming. In fact, this small nation was the second country in the OECD, after Sweden, to institute emission targets (International Energy Agency 1992).

In addition to the fact that the Dutch have an obvious interest in mitigating climate change to reduce future sea-level rise, their position on climate change is consistent with the country's reputation as a leader on other environmental fronts. It is within this background of progressive and environmentally friendly politics that the theory of ecological modernization was developed.[1] In the words of Mol and Sonnenfeld ecological modernization is characterized by "decentralised, flexible and consensual styles of governance," such as that seen in the Netherlands (2000, 6). The similarities between this theory and the Dutch experience itself has resulted in critiques by scholars such as Hannigan (1995, 184), who notes that the theory may not take into consideration "the social and political barriers which are likely to be faced in trying to implement these strategies, especially in countries other than Germany and the Netherlands where the environment is already a major priority."

At the same time, it is unwise to look at the Dutch position on climate change without presenting it in the context of the European Union.[2] Within international negotiations, such as those for the Kyoto Protocol, the members of the union negotiate in a block, with a rotating presidency. As the EU has increased in membership to its present level of fifteen member states[3]—a level of membership that has been consistent since 1995, two years before the Kyoto Protocol was drafted—a decision regarding the union's position on climate change was originally slow to develop. In particular, because of the varied levels of development of the nations of the union, initial attempts to implement measures for CO_2 stabilization were "fiercely resisted" by some member states (Vellinga and Grubb 1992, 2). Perhaps Paterson best describes this situation: "The EU initially went into crisis over problems of implementation of its target, but later picked up momentum to push for reductions" (1996, 74).

However, the European position has since been established, and the EU has subsequently considered the negotiation for a treaty to mitigate global climate change as a "great opportunity to take a leadership role in the UNFCCC process" (Kanie n.d., 11; see also, Huber 1997). In 1996, the EU was one of the first to make the radical statement that it "supported action going beyond 'no regrets'" (Leggett 1999, 236). In other words, the European Union pushed for climate change mitigation measures that would cost money. The union's commitment to reductions can be seen in its "pressure for high targets at Kyoto" (Gummer and Moreland 2000, iv). The Kyoto Protocol, itself, includes the commitment of an EU-wide emission reduction of 8 percent that has been distributed unequally between its member states based on their levels of economic development and their estimated ability to reduce emissions easily. The Dutch have been assigned a 6 percent emission reduction as their share of the EU commitment—a number that falls in the middle of the assignments, which range from a 28 percent emission reduction placed on Luxembourg to the 27 percent emission increase allotted to Portugal. Through this burden sharing, the European Union had come close to stabilizing its emissions at the 1990 levels as early as 1998.[4] Scholars such as Gummer and Moreland (2000) have pointed out that the Netherlands played a leadership role in developing the EU position on climate change, describing the country as having been "at the forefront of EU appeals for action on climate change at both Rio and Kyoto" (Gummer and Moreland 2000, v; see also Kanie n.d.).

Since the establishment of the EU, the Netherlands has been "considered an environmental leader both within the EU and internationally" (Gummer and Moreland 2000, 28). As an active member state, the Netherlands has rotated through the leadership of the union. At the 1997 climate change negotiations in Kyoto, the Dutch held the presidency of the EU, and some say their leadership resulted in a very progressive EU position on emission reductions (see, e.g., Leggett 1999). In the words of Rafe Pomerance, the deputy assistant secretary of state for the United States from 1993 to 1999, who negotiated for the United States in Kyoto, "The Dutch had gotten the EU to agree to minus 15 percent" below 1990 levels (interview, Pomerance 2001).

The positions of the EU member states have significant bearing on the climate change regime for the union as a whole. In fact, the relationship between its member states and the EU provides more support for the global environmental system. Before the EU considers the ratification of the Kyoto Protocol, for example, all fifteen member states must domestically ratify the protocol. Even before the COP-6 negotiations began in The Hague in November 2000, a number of European states, including the Netherlands, had begun the ratification process. In the words of Henriette Bersee, the directorate general for Environmental Protection at the Ministry of Housing, Spatial Planning and the Environment, "We, the Dutch, have started already. We sent our proposal to the department, but it will not be ratified before COP-6. . . . I think France has already sent it to [their] department and probably . . . the United Kingdom" (interview, Bersee 2000).

As a precursor to the ratification of the protocol, the Netherlands developed a plan for meeting its emission reduction requirements of 6 percent below 1990 levels by 2008 to 2012. Carbon dioxide emission reductions, however, are not new to Dutch environmental policy. They have been part of Dutch national environmental policy plans (NEPPs) since 1989 (for a complete discussion, see Kanie n.d.; see also, International Energy Agency 1994). The first NEPP, approved in 1989, set the year 2000 as the target for the stabilization of Dutch CO_2 emissions at 1990 levels. The following year, a subsequent environmental plan, known in the Netherlands as the NEPP-plus, changed emission reduction targets to involve stabilization at the 1990 level by 1995 and added a 3 to 5 percent reduction by 2000. In 1993, a third environmental plan was introduced. The so-called NEPP-II changed the carbon dioxide reductions once again—targeting a 3 percent domestic reduction of CO_2 emissions by 2000. Since 1993, Dutch policy regarding emission reductions has been consistent. If achieved, the 3 percent emission reduction below 1990 levels inside the Netherlands, combined with an additional 3 percent targeted through Dutch-sponsored programs overseas, would meet the Dutch commitment to the EU for a 6 percent total reduction under the Kyoto Protocol.[5]

Environmental performance itself, however, has been another matter altogether. Although the Dutch have been a global leader in supporting and implementing policies to mitigate climate change, not one of the emission reduction goals set by the national environmental policy plans has been achieved. As discussed in fuller detail in the sections that follow, unexpectedly high levels of economic growth, combined with an economic base "geared to high energy use" (Gummer and Moreland 2000, 28; see also, Kanie n.d.), has made the attainment of the emissions targets difficult to achieve. In fact, Dutch emissions in 1998 were 9.3 percent above the 1990 baseline year level and thus 16.3 percent above the country's overall goal (International Energy Agency 2000a). Given these levels and their goal of reducing domestic emissions to 3 percent below the 1990 levels, the Dutch must cut emissions within the Netherlands by 12.7 percent while successfully implementing projects that account for an additional 3 percent of the country's original emission reduction commitments abroad.

This chapter examines this paradox of policy and performance in Dutch responses to the Kyoto Protocol leading up to the final round of negotiations, the Conference of the Parties-6 (COP-6)—of which the first part took place in The Hague in November 2000 and the second in Bonn in July 2001. First, I analyze qualitative interviews of climate change leaders involved in domestic decisions to understand the Dutch position on climate change. Then, I conclude by summarizing the effects of these positions in terms of the formation of the Dutch climate change regime.

RESULTS: POST-KYOTO STATUS

Science

In contrast to Japan, where few questions remain regarding the science of climate change, questions do remain in the Netherlands. Most of the people challenging the validity of the science of climate change in the Netherlands tend to work within the industrial sector. One important view comes from the Confederation of Netherlands Industry and Employers (VNO-NCW), which lobbies the Dutch government on behalf of 95 percent of large businesses and 60 percent of small- and medium-sized enterprises in the Netherlands. Wiel Klerken, the director of environmental affairs for the confederation summarized the general position of many industries in the Netherlands: "In industry circles, there is still a lot of doubt on the issue of climate change and whether it is a problem or not" (interview, 2000). Although there is some skepticism within the Dutch industrial community about the science of the issue, the science is no longer challenged within the public arena.

Still, even with the continued misunderstanding and disbelief in the science of climate change, these opinions affect neither the work of scientific institutions nor the Dutch government's position on climate change. Dr. J. Verbeek of the Royal Netherlands Meteorological Institute (KNMI), which has been involved in presenting the science of climate change to the national government and the public since it became an issue of concern, states that KNMI has not been overtly criticized for its work on the science of climate change, "not at all" (interview, 2000).

The skepticism put forth by actors from the market sector is perhaps best described by Bert Metz, the cochair of the Intergovernmental Panel on Climate Change, Working Group III (IPCC WGIII); the head of the global environmental assessment division at the Dutch National Institute of Public Health and the Environment (RIVM); and the past Dutch delegation leader to the negotiations through the 1997 COP-3 meeting, in Kyoto:

> As soon as they do not like the consequence of scientific findings, they try to question it, and that is exactly why there is a controversy about science and

climate change: because many people don't like the consequences because they do not see it is in their interest. If you look at the IPCC reports, IPCC is trying to make a very balanced assessment. Nevertheless, there are always a few people that stay outside. . . . In many cases they do not want to be involved because they would have to accept the consensus-making process and they keep saying different things, they keep twisting the facts in many cases, or they just play the uncertainty card very much. (interview, 2000)

This description provides one likely interpretation of the lack of support for the science of climate change in industry circles: since climate change mitigation could well cost at least some companies a significant amount of money, it might make tactical sense for some industrial representatives to challenge the validity of the underlying science.

Even with industry showing some level of skepticism of the science, the Netherlands has moved forward in dealing with climate change as an issue. One of the reasons is perhaps best explained by Pier Vellinga, director of the Institute of Environmental Studies at the Free University of Amsterdam and a leader of the Dutch government's climate change program through COP-2, in 1996: "In the Netherlands, we have a lot of these types of relationships, where academics and CEOs and politicians know each other. It's a very small country, and so I guess that's why you get quite a homogeneous opinion of what's going on" (interview, 2000). In his statement, Vellinga alludes to what the Dutch call the *polder model* for negotiation and policy making. Although collaboration is not a new characteristic of the Dutch people, the model, which builds off of ideas put forth by Lijphart (1975), became a particularly well-known method of resolving political issues in the Netherlands in the 1980s, focusing on cooperation and consultation between actors.[6]

Although scientists work with government officials on climate change in the Netherlands, the relationship is more advisory than in Japan. Merrilee Bonney, a member of the directorate general for Environmental Protection at the Ministry of Housing, Spatial Planning, and the Environment (VROM) and one of the authors of the ministry's *Climate Policy Implementation Plan* (Ministry of Housing, Spatial Planning and the Environment 1999a,b), emphasizes this relationship between scientists and policymakers in the Netherlands: "I would have said simply that the scientists are very careful to make clear for the policymakers what is certain, what is uncertain, and how they should . . . deal with the information that is produced" (interview, 2000).

As part of the advisory role that science and scientists play in guiding policy, scientists do maintain distance between themselves and policymakers. For instance, Verbeek explains the KNMI's push to maintain some distance from the regulating bodies of the government: "You have to separate the science, the knowledge side, from the policy side, because there are conflicts. The science has to be objective, you know, and it can be, because we don't do anything in practice. Once you start implementing policies, you have to make judgments and all that, you know; it's not objective any more, and it should not be, because . . . that's how the world is" (interview, 2000). This self-imposed distance between the scientists and policymakers in the Netherlands may permit industrial actors to express their concerns about the science of the issue more vigorously. Nonetheless, this lack of complete consensus has not affected the state's policy-making ability.

Another reason that the Netherlands moved forward in dealing with climate change as an issue is provided by Peter Scholten, the deputy general of Energy for the Ministry of Economic Affairs, who sheds some light on why industry has given up challenging the science of climate change in the Netherlands. He points out that, since Kyoto, the Economic Ministry—the ministry that works most closely with industry—is "interested in the development of the scientific activities in relation to climate change, but for our work we . . . just have this political decision and there is no need within the ministry . . . to discuss if they felt a real danger or not. There is this political decision. . . . There is no need to have this kind of discussion with the ministry until there is new real evidence on the table" (interview, Scholten 2000). In other words, perhaps due to the middle-range relationship between science and policy making, climate change is no longer seen as a scientific question in the Netherlands; it has become a policy question. As this high-ranking government official in the Ministry of Economic Affairs points out, the science and policy of climate change in the Netherlands are presently considered two analytically separate issues. Even the Ministry of Economic Affairs, the arm of the government that works most closely with Dutch industry—the sector of Dutch society that continues to challenge the science of the issue—has accepted the validity of climate change and will not challenge it. Instead, many Dutch industries have identified an economic incentive for supporting climate change mitigation, rather than continue to question the science.

Industry

Prior to the Kyoto round of negotiations, Dutch industry did not actively support the government's climate change policies. In the words of Vellinga,

> I would say industry was lagging behind until Kyoto, and Kyoto brought home, really, the message. Whatever is the problem, at least there is a public opinion so rigorously pursuing regulation measures. . . . That's at least what made Shell move . . . because people want to go to renewables anyway, they

want energy efficiency anyway, and . . . not only Shell, but a few other companies . . . and they suddenly discovered that there may be opportunities for
them. (interview, 2000)

Peter Kwant, the group research advisor for innovation and sustainable
development at Shell International, reinforces Vellinga's summary in his
representation of the prevailing opinion within the industrial sector in
the Netherlands: "The days are over that we would even hope to influence the sort of processes [regarding climate change policy] in a major
way" (interview, 2000). Still, it would be inaccurate to conclude that
Dutch industry has ceased to challenge the science of climate change. At
this point, in the words of Wiel Klerken, the director of environmental affairs for the VNO-NCW, "We do not bring forward that argument [that
the science is inaccurate] so often anymore because it is no use in the debate in this country: It is no use. But that is still the feeling [within in-
dustry]" (interview, 2000). In contrast to the progressive position taken
by its counterpart in Japan, the Keidanren, it may be noteworthy that researchers involved in the government-funded Climate Options for the
Long Term (COOL) program, such as Marleen van de Kerkhof, the project manager of the domestic section of the program, describes the VNO-
NCW as being less forward-thinking on the issue than the companies
that it represents (interview, 2000).

In the same vein, other spokespeople for industry—including Gert van
Ingen, the president of Akzo Nobel Energy, the energy group of one of the
leading chemical companies in the Netherlands—were less consistent in
their support for the science of the issue. In van Ingen's view, "there is no
doubt" about the human influence on climate change. As he put it, "Carbon dioxide in itself is a serious matter" (interview, 2000). Still, although
the company states that they recognize the significance of climate change,
the president later challenged the statement in the interview: "This enhanced greenhouse effect due to carbon dioxide is, in itself, bah, ridicu-
lous. Again climate change, there is no doubt. . . . That's for sure, but the
emphasis on carbon dioxide and its effect on climate and the so-called evidence of that . . . does not meet the scientific quality that is required for
real science" (interview, 2000). In other words, although the company officially acknowledges the importance of the issue, people working at
Akzo Nobel Energy, including the president, do not see how anthropogenic CO_2 emissions contribute to climate change. This lack of understanding of the science of climate change put forth by members of the
market sector is not a surprise.

Although some members of the market sector continue to challenge the
science of climate change, most of them do not—including other energy
companies. Shell International, one of the largest oil companies in the

world, provides an example of a company that has taken a rather progressive position in dealing with climate change: the company has been investing in alternative energy and other climate change mitigation measures since 1997. In the words of Kwant, "The climate change issue was discussed and we found that, yes, of course, it is sensible to prepare yourself in case society might need these sort of responses" (interview, 2000). Here, Kwant cites the precautionary principle as Shell's reason for investing in alternative energies. If scientific projections are correct, Shell's position should be quite profitable for the company. Although this position is relatively consistent with that of British Petroleum (BP), it contrasts markedly with the positions of U.S. oil companies such as Exxon and Mobil, which continue to challenge the idea that climate change is taking place at all.[7] It is important to note that Kwant's statement is consistent with the explanation of the precautionary principle put forth by van Tatenhove, Arts, and Leroy, who claim that "in the Netherlands, the implementation of the precautionary principle was part of the . . . discourse and target group policy. . . . New responsibilities were formulated for involved parties, resulting in a shift from direct regulation to market self-regulation and experiments with economic instruments" (2000, 3). This implementation of the precautionary principle helps to explain the role that Dutch industry is playing in climate change mitigation.

For the most part, that role has been forward-thinking. Most industrial actors in the Netherlands are not fighting national climate change policy; they are responding with action, identifying economic opportunities in climate change mitigation. Many of the industries in the Netherlands now perceive their economic self-interest in ways that may seem unusual to readers who are more familiar with U.S. companies' responses (discussed in chapter 6). Marjolein Quené, the manager of business development for the international and renewable energy section of NUON, the company that provides 40 percent of Dutch energy, describes her company's motivation:

> We want to be a front-runner in renewable energy. That's really the policy of the company, not only from this division, but from the company as a whole, and we are [the] front-runner in the Netherlands. We want to be one of the biggest sellers of renewable energy in Europe, so we are really investing in that, and we see there is a future for renewable energy also from a commercial perspective." (interview, 2000

Similarly, Shell International's motivation has been described as an attempt "to be more a part of any solution to the problem" (interview, Kwant 2000). The types of industrial innovation that are described by Quené and alluded to by Kwant are consistent with the "superindustrialization" predicted by the theory of ecological modernization.

With industry's identification of market opportunities, there has been a consistent move among companies in the Netherlands to address the issue of climate change. The deputy director general at the Ministry of Economic Affairs points out that, even with some of the skepticism remaining regarding the science of climate change, Dutch industry has taken an active role in responding to the issue: "More and more . . . international companies are already, in their internal business, dealing with the carbon dioxide greenhouse emissions . . . [at] BP and Shell, but also a lot of other companies, they already have [their] own carbon dioxide targets, they are dealing already with this. . . . When you add the list of big companies, the oil companies, the financial institutions, they are investing quite some money all expecting that carbon dioxide will be a relevant issue and that [the investment] will have some value" (interview, Scholten 2000).

A case in point is provided by the fact that most industries in the Netherlands have actually endorsed the Kyoto Protocol. In the words of Klerken,

> We have signed the protocol. VNO-NCW signed it and then trade associations, industries have signed it. The chemical industry and mace metal industry . . . electricity plants, paper mills, etc. They have signed the agreement and now the individual companies are signing, they are in the process of signing, and I think at this moment, 80 percent of the companies which could sign have signed, which in effect covers also 80 percent of industrial energy use. They all signed. (interview, 2000)

Although industries' endorsement of the protocol through their signing has no official role within the Dutch ratification process, their support was an important step in beginning to ratify and implement the agreement.

One of the predominant methods for Dutch industry to respond to the issue of climate change is through voluntary agreements with the state—an approach endorsed by Klerken: These agreements "have been very successful because they give a flexibility to companies to solve their environmental problems. . . .We make an agreement on solving the problem within a certain period of time and they are free to choose how and when they solve the problem, which gives them flexibility to make smart combinations with other investments they are already planning within their firm" (interview, 2000). Perhaps the most well-known examples of voluntary agreements between actors are the energy efficiency covenants and the industrial benchmarking plan, through which companies commit to monitoring their energy efficiency: "In return, the government will refrain from other measures towards these industries" (interview, 2000).

The energy efficiency covenants, in fact, began as early as 1989. The goal of these covenants between the Dutch state and industry was not emission reductions per se but rather improvements in industrial energy efficiency. The agreements were voluntary, and they maintained the firms' privacy by employing a third party to monitor progress. In addition, these covenants are also extensive: "Long-term agreements covering the period 1989 to 2000 were concluded between the Ministry of Economic Affairs and nearly all industrial sectors with substantial energy consumption. In view of their voluntary character and the high degree of public–private cooperation," these agreements constitute a major Dutch policy decision (Enevoldsen 2000, 78).

Another example is provided by the benchmarking plan. Within this plan, industry is autonomous in deciding how it will respond to the Kyoto Protocol. Not surprisingly, the managers of companies have declared that, rather than be regulated, they would prefer to have voluntary agreements with the state, including the autonomy to decide for themselves how to achieve the necessary results. Industry spokespeople such as Klerken made it particularly clear that the fear of regulation and taxes has motivated Dutch industry's response to climate change. In his own words, "That is the only reason" (interview, 2000). Clearly, the pressure derived from the public, along with the fear of government regulation, played a role in the VNO-NCW's position.

As in the case of Japan, however, there is some question as to the efficacy of such voluntary measures. In particular, given the continued increases in Dutch greenhouse gas emissions in recent years, some experts are beginning to question whether the Kyoto Protocol targets can be met without state-led regulations. Metz spoke about the future of voluntary agreements: "Because the required reductions are such that these voluntary agreements do not deliver enough any more, I think it is a matter of timing. The required reductions . . . go faster than the voluntary agreements basically can deliver. So I would think the time for voluntary agreements on climate has gone" (interview, 2000).

Even some industry members agreed with Metz. In the words of Quené, "I think that sooner or later, they will have to [have binding agreements]. . . . Now it is a voluntary agreement. . . . Why not the obligation starting now [by saying that] 3 percent of the energy . . . should be renewable. . . . See, then, everybody knows what they have to deal with. You are certain you will reach your goals at least for electricity and energy, and if you start this process now, it is very clear" (interview, 2000).

Given the significant emission reductions required by the Kyoto Protocol—and the fact that Dutch emissions have increased to over 16 percent above the goals of the first commitment period since the protocol was drafted—climate change policy in the Netherlands, even with the voluntary agreements, will have to change if the Dutch are to meet their target. Whether the new policies will be driven by industry or social actors outside of the industrial sector has yet to be seen. Thus far, however, the state has been industry friendly and has supported Dutch industries' development of their own efforts to deal with climate change.

The State

Given the Dutch government's acceptance of the science of climate change and its policy decision to treat global warming seriously, the state has played a significant role in the development of the climate change regime in the Netherlands. Beyond taking the advice of scientists, the Dutch state has also worked to collaborate with other social actors in its work on climate change. As noted earlier, the Dutch parliament had started its domestic ratification process even before the COP-6 negotiations. In fact, the week before the COP-6 negotiations began, the Dutch parliament debated the issue of ratification of the Kyoto Protocol. In the words of the chair of the committee for global climate change in the Second Chamber of Parliament, Eimert van Middelkoop, the goal was to "make a beginning before COP-6" (interview, 2000). The original intention of the parliament was to conclude discussions about ratification of the protocol directly after COP-6 ended, but due to the stalling of the negotiations on the last day of the international meeting in The Hague, the parliament delayed any further discussion of ratification until after the second part of the COP-6 negotiations concluded, in July 2001.

Even with these unexpected delays in the ratification process, those involved in the formulation of the Dutch climate change regime agree that the Dutch will eventually ratify the Kyoto Protocol. Industry spokespeople such as Hans Davidse, the manager of technology and energy conservation at Akzo Nobel Energy, said that "the political momentum is so great that maybe the signing of the Kyoto Protocol is more of a formality than a practice. The whole political movement is going on as if Kyoto has been signed, so in practice . . . it is really on its way"[8] (interview, 2000). With business leaders resigned to the eventual regulation of climate change in the Netherlands, political discussions about global warming moved from "whether" to "how." In other words, the discussions did not debate the future of the protocol; rather, they focused on what would bring about the emission reductions that are necessary to meet the Kyoto commitments. In the words of Metz and colleagues (2000, 6), the EU has been "building trust with key countries for getting the necessary support for entry into force of the Kyoto Protocol without the USA."

The fact that industry is prepared for, and not trying to stop, the ratification process is consistent with the Dutch polder model, in that the state is working closely with industry and civil society actors. In other words, the case of the Dutch climate change regime represents what Lijphart

(1975) would call the "politics of accommodation," in that the state and industry are consulting with one another and producing a progressive climate change regime for the Netherlands, with the goal of meeting the Dutch obligation to the Kyoto Protocol.

As mentioned, the Dutch government's national environmental policy plans have included CO_2 emission reduction targets since 1989, well before the Kyoto Protocol was even drafted. Still, although these plans have long existed, the targets have never been met. In the words of Metz, "There were 1994–1995 goals and there were 2000 goals. None of them has ever been achieved. . . . None of that worked" (interview, 2000). Sible Schöne, manager of the climate change programme for the Dutch World Wide Fund for Nature (WWF) and one of the environmental leaders for the issue in the Netherlands, went into further detail about why the NEPP repeatedly failed:

> I sometimes call . . . [the Netherlands] a progressive OPEC country. . . . The Dutch economy is heavily based on our gas reserves, and thanks to the gas reserves and the fact that Rotterdam is the biggest harbor in the world, the Netherlands is attracting energy-intensive industries from everywhere. (interview, 2000)

As Schöne points out, economic growth in the Netherlands, particularly in the energy-intensive sector, had resulted in significant increases in CO_2 emissions, even during the period when environmental policies were in place to reduce these emissions, in part because the Netherlands is home to more energy-intensive industries than other members of the EU.

Increases in CO_2 emissions during periods of economic growth, of course, can be seen throughout the world. In the words Klerken, "Economic growth is the problem. There's almost a one-to-one relationship between economic growth and energy use . . . that makes it very difficult. . . . Everything you do to save energy blows away with such an economic growth. . . . Our economic growth at the moment, we are very happy with that of course, but it's a problem" (interview, 2000). Klerken does appear to have overstated the relationship between economic and energy use in the Netherlands; research into this relationship shows that, in the Netherlands, "the economy—in terms of GDP—grew faster . . . than the energy intensity decreased" (Farla and Blok 2000, 114). This disconnect between economic and energy intensity suggests that the Dutch government is indeed trying.

As part of that effort, and in light of earlier failures to meet the NEPP emission reduction targets, the government drafted a new environmental plan in 1999 that was specifically designed to address the issue of climate change: *The Netherlands' Climate Policy Implementation Plan, Part I:*

Measures in the Netherlands (Ministry of Housing, Spatial Planning, and the Environment 1999a). The plan follows through with the goals of the earlier NEPPs and aims to achieve the Dutch emission reductions outlined in the yet-to-be-ratified Kyoto Protocol. The climate plan was adopted by the parliament in November 1999. Part I of the plan focuses on domestic actions that are designed to achieve the 12.3 percent emission reductions that are to be achieved inside the Netherlands. With this 12.3 percent reduction, Dutch emissions will achieve the 3 percent domestic reduction below 1990 levels. To ensure that the 3 percent domestic reduction targets are met with this domestic plan, "it has these evaluation moments built in for 2002 and 2005" (interview, Bonney 2000; for a full discussion of the implementation plan, see Ministry of Housing, Spatial Planning, and the Environment 1999a, 1999b). The other 3 percent of the emission reductions are intended to be carried out internationally and are outlined in the second part of the plan, *The Netherlands' Climate Policy Implementation Plan, Part II: The Context of Climate Policy* (Ministry of Housing, Spatial Planning, and the Environment 1999b). Although some industrial actors think that "the Netherlands will not be able to meet this 6 percent [below 1990 level] reduction target" (interview, Klerken 2000), most claim that the targets will be met by the 2008–2012 commitment period.

One of the authors of the domestic part of the climate plan described the different groups involved in meeting its goals: "A lot of different people are now responsible for implementing those measures because they relate to traffic, to energy, to industry, to households, to renewable energy, but the overall sort of coordination is from this office. We make the progress reports for the quarter for the minister. We report to Parliament about everything " (interview, Bonney 2000). Unlike the previous emission reduction strategies included in the NEPPs, the Dutch climate plan is designed to be adjustable to ensure that the goals are met within the stipulated period.

The governmental policies, moreover, now go beyond collaborating and reporting. In addition to continuing earlier environmental policies that focus on energy, such as a fuel tax and an assortment of incentives for fuel-efficient automobiles, the new plan emphasizes what may be one of the most successful policies to date, namely, a small-scale energy consumption tax that was first implemented in 1996. It was called a "substantial energy tax" by environmental leaders such as Schöne (interview, 2000). Jan Pronk, the minister of the Ministry of Housing, Spatial Planning, and the Environment, described the tax in a speech he gave at the Conference on Innovative Policy Solutions to Global Climate Change on April 25, 2000: "This tax has raised the gas and electricity prices paid by households and small businesses by about 50 percent. These increases have improved the market position of renewable energy considerably, since renewables are exempt from the tax" (2000, 11). In other words, only fossil-fuel-based energy consumption is taxed. As a result of this policy, "wind energy is competitive with fossil fuels" (interview, Scholten 2000). The policy was designed to support the development of renewable energy and to try to "influence household energy use" (interview, Bonney 2000).

Still, although it is too soon to determine if this tax will be effective at influencing households, it is possible to make one observation with certainty: the new tax does not extend to the large energy consumers responsible for the majority of greenhouse gas emissions in the Netherlands—industry. Instead, in the words of Andre van Amstel, professor of environmental sciences at Wageningen University and an emissions inventory expert for the Dutch government, the "bigger companies and even the biggest companies . . . are getting all kinds of benefits" at this point (interview, 2000). Van Amstel went on to explain why Dutch industries are not taxed at the same level: "We can't impose tax on industry while they can threaten our government that they will leave the country and go to the next. So we need this European level energy tax so that companies can only threaten to go to another continent" (interview, 2000). Due to the autonomy of Dutch industry, the government is cautious about taxing them for their energy consumption.

Even though the EU-wide energy tax was mentioned as a possible opportunity for extending Dutch energy policy to industry, there is significant disagreement among actors in the Netherlands about the efficacy and implementability of such a policy. First, Scholten states that "there must be policies and measures on the European scale" (interview, 2000). However, Metz says that the possibility of an EU-wide green tax is "out of the question. There have been attempts for ten years now to implement an EU energy tax and so far that failed" (interview, 2000). The final decision regarding an EU-wide energy tax will be determined when the EU moves to ratify the Kyoto Protocol—a step that is contingent on all EU member states ratifying the protocol themselves. At the same time, scholars such as Arthur Mol of Wageningen University suggest that the tax may be possible. In his own words, "Several countries have already started an energy tax and the Netherlands will probably join a front runner group if that is large enough" (correspondence with author, July 2001).

One of the other dominant climate change policies in the Netherlands is a benchmarking plan for industry. This plan deals with "energy efficiency" and exempts companies from further emission reduction requirements if they are 5 percent over the efficiency of companies in their respective fields. Minister Pronk also discussed this plan in his speech on climate change.

> Large, energy consuming companies have entered into an official agreement with the government called the Benchmarking Protocol. They have committed themselves to becoming among the most energy efficient industries in the world by 2012, which will help limit the growth in CO_2 emissions. We also

agreed that the benchmark—"who are the most efficient industries in the world?"—will not be defined by industry themselves, nor by the government, but by the two together, assisted by independent experts, according to agreed procedures. (2000, 11)

The program was summarized in the words of van Amstel, "If a company is the most energy efficient in the world, you should not ask for more" (interview, 2000).

Not only does this statement by the environment minister support the Dutch polder model of consultation, but it also is consistent with many of the expectations identified by ecological modernization theorists who predict that the state will make "unproblematic use of science and technology in controlling environmental problems" (Mol and Spaargaren 1993, 454). In general, the Dutch state is playing a middle-range role, between progressive policy and collaboration with the other actors that are involved in the development of the climate change regime in the Netherlands. The general consensus among actors in the Netherlands is that, although the state will support climate change mitigation, the major changes will come from other actors. Vellinga best summarizes this position: "In the end it will not be government pushing it. It will be business opportunities and the public, so it will be civil society and the private sector. Civil society has political power, but civil society also is consumers" (interview, 2000).

Civil Society

Although Vellinga predicts a strong role for Dutch civil society in formulating the climate change regime, the role that it is currently playing is not at the level of leadership that we see with Dutch industry. In fact, it is difficult to identify and characterize the role of civil society in the Netherlands. Actually, people employed by nongovernmental organizations (NGOs) who work on the issue disagree; perhaps Hans Altevogt, the head of the climate change and energy program at Greenpeace Netherlands, best sums up the opinion: "Right now there is not much of a public voice. We have tried, and not only we, but a lot of other NGOs, have tried to organize, but it is very difficult. And it has everything to do with the difficulty of the issue of climate change itself" (interview, 2000).

Altevogt is not the only one to attribute the lack of civil society involvement in the development of the climate change regime to the lack of public understanding and the difficulty of the issue. Metz saw the issue similarly. In his own words, public "knowledge about climate issues is not particularly good" (interview, 2000). Similarly, industry representatives such as Wilhelma Kip, the executive officer of European affairs for EnergieNed, the federation of energy companies in the Netherlands, says that "the general public does not know what it's all about" (interview, 2000). In short, the consensus of those working on climate change in the Netherlands is that the Dutch citizenry does not really understand the phenomenon.

Still, perhaps civil society's role in the development of a climate change regime in the Netherlands is difficult to identify because of the many types of civic engagement that have been conflated under the general term "civil society." One of the major ways that civil society is expected to participate in an advanced industrial nation's public sphere is by participating in mediating groups, such as professional social movement organizations (SMOs).[9] Through such organizations, citizens are said to choose political tactics, such as lobbying and special projects, to voice their preferences (Zald and McCarthy 1972; see also Lofland 1996; McCarthy and Zald 1977). In the case of this type of citizen action surrounding the Dutch climate change regime, a representative from one such SMO stated, "We address the issue to the public and to the citizen but, really, the role of the citizen has not been organized yet" (interview, Altevogt 2000).

Although Altevogt does not find much citizen engagement in his work at Greenpeace Netherlands, other Dutch SMOs have been more successful. For example, the World Wide Fund for Nature (WWF), the sponsor of one of the biggest international NGO teams working on global climate change, recently sponsored a campaign in the Netherlands to educate citizens about climate change through green electricity. The manager of the climate change program, Sible Schöne, described the campaign this way: "It says you can do something, you can help tackle climate change. We don't ask impossible things from you, you can just buy green electricity. And we got fifty thousand new customers in a few months" (interview, 2000). Perhaps the reason that this WWF project was so successful in engaging citizens is that it appealed to an altogether different type of civic action—that of civil society acting as consumers and expressing interest and concern for social issues, such as the environment, with their pocketbooks.

The most obvious expression of civil society interest in the issue of global climate change can be seen in their support of the green-energy tax. Vellinga summarizes citizen support for this tax: "In the Netherlands, they pay more for green electricity than for black electricity on a voluntary basis" (interview, 2000). Other lifestyle changes are less obvious but still important to note. "Citizens have done the most, I think. We have had massive programs for installation of houses, and older constructions, so we have double-glazed windows, we have insulating panels between the two walls of houses" (interview, van Amstel 2000). In other words, citizens in the Netherlands have voluntarily opted to pay a higher cost for their electricity, thereby supporting governmental and industrial opportunities to develop clean energy alternatives. On top of supporting the small consumer energy tax and taking measures in their houses to be

more energy efficient, Dutch citizens express what scholars such as Ingel-
hart (1995) would call "postmaterialist values":[10] they bring shopping
bags with them when they shop; they insist on working shorter hours
than their counterparts outside of Europe; and, in many cases, Dutch citizens ride their bicycles, walk, and take public transportation over driving their own automobiles.

The Dutch climate change regime, in fact, may provide a clear case of
civil society working "'in between' the state and economy" (Emirbayer
and Sheller 1999, 151; see also, Wuthnow 1991). The example of the
citizen-supported green-energy tax is a case in point. In other words,
civil society is not absent from the climate change regime in the Netherlands; it just does not express itself predominantly through SMOs such
as WWF and Greenpeace. Instead, it tends to make itself heard more often as consumers that are internal to the decision-making processes of
the society, rather than as activists.

This line of argument, however, is not without its limitations. Perhaps
most notably, although Dutch citizens have expressed their postmaterialist
values in a number of consumer actions having to do with purchasing decisions, their consumption levels continue to grow. The number of home
appliances and automobiles has significantly increased with the prosperity of the recent economic boom. In addition, as pointed out by Bart Thorborg, of the Department of Strategic Planning at the Ministry of Transport,
Public Works, and Water Management, "the tendency is bigger cars . . .
[and] more cars driving. . . . [Recently,] all emissions dropped very steeply
except carbon dioxide" (interview, 2000). Perhaps the Dutch case highlights one of the major paradoxes of postmaterialism: even if citizens are
willing and interested in purchasing products that are environmentally
friendly and more expensive, their high standards of living mean that they
still can purchase more products and that they may choose to do so.

Nonetheless, the fact that consumers have been voting with their pocketbooks may play a role in this scenario, in that representatives of SMOs
have been included and involved in the policy discussions and decisions
regarding climate change mitigation measures in the Netherlands since the
beginning. Some scholars have attributed this fact to characteristics of the
political culture of the Netherlands. Van Tatenhove, Arts, and Leroy, for example, state that "in the Netherlands, the environmental movement has a
quite different position. In the consensus tradition of Dutch policy making
it is frequently consulted about many environmental issues" (2000, 5).

Beyond the Dutch polder model, consumer actions have given SMOs
more strength at the bargaining table and may contribute to Dutch environmental groups' ability to be internal to the national decision-making
process and take part in policy discussions. Even Greenpeace, which is
against certain aspects of the Dutch position on the Kyoto Protocol, was

invited to and participates in policy discussions. In the words of Altevogt, "We are involved in the talks, apart from one time when I really did not show up out of protest" (interview, Altevogt 2000). In general, NGO leaders work with the Dutch government to help form a consensual Dutch climate change regime. Although difficult to identify, it is obvious that civil society in the Netherlands is contributing to its formation. It may not be as overt nor as interested in advertising its actions as Dutch industry or the state, but it is relatively clear that, if Dutch civil society were not playing multiple roles between the state and the market, the climate change regime in the Netherlands would be much weaker.

CONCLUSION

At the heart of climate policy in the Netherlands is the collaborative style of politics that can be seen in all policy circles in this country: the polder model. In the words of Joyeeta Gupta, a well-published scholar of climate change who works at the Free University of Amsterdam's Institute for Environmental Studies, "This is a consensus society. All actors work together" (interview, 2000). Due to the consensual character of Dutch society, civil society plays an internal role by working with an autonomous market sector, with a state of medium strength, and with scientists that are relatively central to bring about more forward-thinking practices and policies.

However, although the Dutch climate change regime provides an example of collaboration and consensus, it has not effectively responded, to date, to the issue of climate change in the Netherlands. Instead, emissions continue to rise. In many ways, the climate change regime in the Netherlands is being led by the state in consultation with industry and civil society—a top-down approach to climate change—but their efforts to reduce the greenhouse gas emissions that are at the heart of the Kyoto Protocol have not been successful. In short, the only actor in the Netherlands that has taken on any burden associated with climate change is civil society—and it cannot reduce emissions alone.

With the change in the U.S. position and the threat of the Japanese pulling out of its commitment to the protocol prior to the COP-6bis meeting, the issue of ratification reemerged in the Dutch parliament in early July 2001. In the words of Jeff Prins, the assistant to the Labor Party in the Second Chamber of the Parliament,

> The conservative VVD party might be recalcitrant on the issue, demanding U.S. participation before committing to any more Dutch actions . . . and, in light of this, politics in the Netherlands is getting exciting. Next year there are elections, so you see more political polarization here. . . . Ratification for the

Dutch will not necessarily be dependent on the U.S. and Japan. Still, Japan is crucial, since it can bring the Kyoto partners past the 58 percent mark [which will meet the requirements of article 25 of the protocol]. . . . Only Bonn can tell what will happen. (correspondence with author, July 2001)

Beyond the question of ratification, however, it is still unclear how the Dutch will achieve the goals put forth by the protocol, since the methods used thus far have been relatively ineffective. The heightened role of the state, as written into *The Netherlands' Climate Policy Implementation Plan: Part I* (Ministry of Housing, Spatial Planning, and the Environment 1999a), will be responsible for enforcement of reductions and thus will take some power away from Dutch industry. The outcome of this new plan has yet to be observed.

On an empirical level, the Netherlands may indeed provide a clear example of a nation-state that has mechanisms taking place as predicted by ecological modernization; however, the environmental outcomes predicted by the theory have yet to come true. Arguably, this small, affluent nation-state may provide a clearer case of citizens acting in ways that are predicted by scholars working within the postmaterialist perspective. Overall, the experience of the Dutch climate change regime supports the idea that, although ecological modernization and postmaterialism are possible, they have not yet substantially affected actual material environmental outcomes, such as emissions. With the Dutch case, we can see another example of how the interrelations among actors within the global environmental system determine the political outcomes of the domestic climate change regime.

NOTES

This research was made possible through the support of the U.S. National Science Foundation and Wageningen University. I would also like to thank the University of Amsterdam, Department of Political Sciences for their donation of office space.

1. Although ecological modernization was originally presented in the German language by Huber (1985, 1991), its most active proponents in the English-language literature, Mol and Spaargaren, are academics in the Department of Environmental Sociology at Wageningen University, in the Netherlands.

2. The European Community became the European Union in 1993; for the sake of brevity, I have used the term European Union (EU) throughout the chapter.

3. On May 1, 2004, ten new member states are schedule to accede formally to the union (Council of the European Union, "Ten New Member States Set to Join the European Union," press release, 2003, available at www.delcyp.cec.eu.int/en/news/16April/tenmembers.pdf [accessed October 2, 2003]).

4. It is important to note, however, that most of these reductions have been attributed to the updating of plants in the former East Germany and an almost complete cessation of coal consumption in the United Kingdom.

5. It is important to note here that the 3-percent-plus-3-percent decision regarding Dutch emissions was not reached without difficulty; instead, it involved significant discussions among social actors within the Netherlands.

6. The term comes from the polder, or the land reclaimed from the sea, and it alludes to the fact that throughout history members of Dutch society have worked together to build and maintain dikes that protect the polders. For a complete discussion of the polder model, see the European Foundation for the Improvement of Living and Working Conditions, "Wave of Reorganisations at Major Dutch Groups: An End to the Polder Model?" *European Industrial Relations Observatory Online*, 2003, available at www.eiro.eurofound.ie/1998/11/features/NL9811106F.html (accessed October 2, 2003); see also NRC Webpagina, "The Political Branch of the Polder Model," *Profiel the Netherlands*, 1999, available at www.nrc.nl/W2/Lab/Profiel/Netherlands/politics.html (accessed October 2, 2003).

7. See www.campaignexxonmobil.org (accessed October 2, 2003) for more information.

8. In this case, when Davidse says "signing," he is referring to the Dutch ratification of the Kyoto Protocol.

9. Although the term SMO usually refers to all types of social movement organizations—both professional and not—for the sake of brevity, I use the term to refer specifically to professional organizations.

10. See also Abramson (1997); Brechin and Kempton (1994, 1997); Dunlap and Mertig (1997); Inglehart (1990); Kidd and Lee (1997a, 1997b); Pierce (1997).

6

Debate and Discord in the United States

In comparison to the Dutch and Japanese responses to the issue of climate change, it is likely that the United States provides the most interesting case of domestic actors competing for control. In fact, since negotiations toward a climate change treaty began in 1995, with the goals of the Framework Convention on Climate Change (UNFCCC) in mind, even the different branches of the U.S. government have been in disagreement about what position the country should adopt in the global arena. Due to the extensive disagreement within the United States about the issue of climate change, as well as disagreements regarding the role that the country should play in the formation of a global climate change regime, the history of the climate change regime in the United States is one of debate and discord.

First, the Clinton administration worked throughout its eight years with the international community to come up with a ratifiable form of the Kyoto Protocol that, as recently as 2000, was still considered, in the words of David Gardiner, deputy chairman of the Clinton administration's White House Climate Change Task Force, "a work in progress. . . . We've been continuing since Kyoto to try to put the agreement in a shape that we feel would be necessary in order for us to submit it to the Senate for ratification" (interview, 2000). However, prior to the Kyoto round of the negotiations, where the protocol was drafted, the Senate unanimously passed a resolution stating that, in the words of a congressional research service report, "the United States should not approve any Kyoto agreement that did not impose binding reduction requirements on all nations and requested the [Clinton] Administration to analyze the costs implied by any treaty submitted for its approval" (Parker and Blodgett 1999, 2). In other words, as was stated in 2000 by a top staff member for a senator who was heavily involved in the climate change debate, the Kyoto Protocol is a "frozen policy" (interview, Foreign Relations Counsel 2000).

The U.S. case does not fit the predictions of any of the environmental state theories that were explored in the beginning of this book. In fact, the results of this chapter may provide some explanation for why much of the research on society–environment relationships in the United States continues to have more pessimistic expectations than does the work coming out of Europe. In particular, the main themes within the debate about climate change in the United States revolve around the issues of developing country participation, binding targets, and the cost of implementation. Perhaps one of the most significant effects that this debate has had in the United States is that interpretations of the science of climate change and the feasibility of the Kyoto Protocol have become polarized. For the case of the United States, this study tries to explain how, in March 2001, while most Annex I nations were intending to begin the ratification process for the Kyoto Protocol by 2002, U.S. president George W. Bush completely changed his administration's position on climate change, sending a letter to Senate leaders stating "As you know, I oppose the Kyoto Protocol."[1]

President Bush's new position on the Kyoto Protocol, although a shock to the international community, was not a surprise to many in the United States. Even though Bush had included industrial carbon dioxide emission reductions in his campaign promises, his prior environmental record as governor of Texas and his connection to the energy industry made his critics skeptical of his following through on his campaign promise. In fact, the U.S. rejection of the Kyoto Protocol had been rumored since well before Bush's inauguration, and a number of popular articles and books have focused on this point using titles such as *The Collapse of the Kyoto Protocol and the Struggle to Slow Global Warming* (Victor 2001; see also Gelbspan 2000; Judis 1999; Sarewitz and Pielke 2000). Still, even though there had been a good deal of speculation that the United States would pull out of the Kyoto Protocol, many high-level people involved in the issue of climate change did not see the move coming. Included among those was Jonathan Pershing, the former U.S. climate change negotiator who became the division head of the Energy and Environment Division of the International Energy Agency, who remarked in December 2000 on the "clear carbon dioxide reductions" included in the energy plan of presidential candidate George W. Bush, suggesting that if Bush were elected, his administration would regulate climate change (interview, 2000).

Unlike other accounts of the climate change debate that have looked at official statements and at the implications of government positions on the global arena,[2] this chapter traces the development of the climate change debate within the United States by following the claims of different social actors and the roles that they have played domestically. Accordingly, I provide analysis of qualitative interviews with U.S. climate change leaders from different sectors involved in the domestic decision-making process. The case study beings with a discussion of the role of industries and the industry-funded think tanks in relation to American decisions about climate change, building on prominent existing work on the subject. Second, I address the role that the science of the issue has played within the climate change debate in the United States, paying particular attention to ways in which the science has been constructed. Third, I look at the climate change debate that has taken place within sectors of the U.S. state, a debate that has been going on almost constantly since the United Nations Framework Convention on Climate Change was signed by the United States in 1992. Unlike my other case studies, this chapter includes a summary of the important political occurrences prior to COP-3, where the Kyoto Protocol was drafted, as they are important to understanding the extreme polarization that existed between the Clinton administration and the Republican-led Congress. This polarization halted any progress toward the development of a climate change regime in the United States since 1997. Fourth, I present the role that civil society actors, including environmental groups, played in the U.S. debate, leading up to the two sessions of the Conference of the Parties-6 (COP-6), the first of which took place in The Hague in November 2000 and the second in Bonn in July 2001. Finally, I conclude with a discussion of the changes in the U.S. position—beginning with President Bush changing his position on the carbon dioxide reductions, followed by Senator Jeffords leaving the Republican Party, thereby shifting the U.S. Senate leadership to the Democrats. Within the conclusion, I discuss what implications these changes in position will have on the possible formation of a U.S. climate change regime in the near future, as well as their implications on progress toward an international climate change regime.

By following the outline described here, I explain how the United States provides a clear and solid case for a materialist interpretation of the differences in national responses to the issue of climate change—a case that goes beyond the divergence within the state and the actors influencing it. In particular, the United States is one of the most automobile- and coal-dependent of the Annex I nations that are party to the Kyoto Protocol. Both in terms of consumption and indigenous resource reserves, this reliance on automobile travel and the extraction and consumption of coal is significant. Related to its dependence on these goods, the United States also presents an example of the success of conservative nonprofit organizations playing a role in contributing to public discourse about the construction of the nation's position.

RESULTS: POST-KYOTO STATUS

Industry

In their article on the conservative movement's reframing of the issue of climate change in the United States, McCright and Dunlap conclude that:

> The controversy over global warming—and the resulting difficulty its advocates have in keeping it on the public agenda—is not simply a function of waning media attention, the ambiguities of climate change signals, or the complexities of climate science, but stems in large part from the concerted efforts of a powerful countermovement. . . . We see that the conservative movement employs counter-claims that serve to block any proposed action on global warming that challenges its interests. (2000, 519–20)

Consistent with the authors' conclusions, many of the American industries that stand to lose money and business have mobilized anti-Kyoto action through the types of organizations studied by McCright and Dunlap. Of these political organizations, which call themselves "conservative nonprofit organizations," perhaps the most well-known is the Global Climate Coalition, which at one point included in its membership multinational corporations such as the Atlantic Richfield Coal and Oil Company, Chevron, Chrysler, Ford Motor Company, Texaco, and the Western Fuels Association. John Passacantando, the executive director of Greenpeace USA and former executive director of Ozone Action, described the original mission of the Global Climate Coalition: "The Global Climate Coalition, representing the auto companies, the coal, car, oil, utility companies, made an attempt to kill the Kyoto Protocol. . . . These are companies that are trying to stop any . . . legally binding international agreement to stop global warming" (interview, 2000). The coalition, which was founded in 1989, lobbies the U.S. government about its position on the regulation of greenhouse gases, domestically as well as at the international negotiations.

This coalition, along with other conservative nonprofit organizations—such as the Greening Earth Society, the Competitive Enterprise Institute, and the National Consumer Coalition (which has coordinated the Cooler Heads Coalition)—do not want the United States to sign on to the Kyoto

Protocol. In contrast to the VNO-NCW, in the Netherlands; and the Kei-danren, in Japan, both of which represent economic interests and trade associations in their own countries, these American organizations identify themselves as nongovernmental organizations (NGOs)—even though they are financially supported, for the most part, by industry itself—thereby gaining access to the media and the public that they may not otherwise have. Glenn Kelly, the executive director of the Global Climate Coalition, spoke about the rationale behind, and the purpose of, these organizations:

> Other governments respect the role that business can and should play . . . in international negotiations, but there's this basic understanding that NGOs are only environmental organizations, so when we're overseas and we meet people for the first time, and say we're the Global Climate Coalition, we're an NGO, the immediate assumption is well, then you're an environmental activist organization. . . . We represent the business community but thank you, I do consider myself an environmentalist. . . . The business NGOs that we work with range the gamut. The North American business NGOs tend to be much more inclined to question the process that is undertaken, they question the underpinnings of the whole . . . whereas the business NGOs from Europe are less willing to come forward and question their government. They don't have that tradition that we have so beautifully ingrained in our Constitution of the right to petition the government and, as a result, I think they are much more cautious and are seen as somewhat less strident. . . . That's sort of the way they view us, but that's just part of our national heritage. It's in our Constitution and that's something that is in our culture. (interview, 2000)

As Kelly outlined, the conservative movement has used some of the major tenets of political culture in the United States to identify its mission, and the rhetoric of individualism and personal rights is an ongoing theme within the movement's anti-Kyoto campaign.

Beyond the Global Climate Coalition, another conservative nonprofit organization is the Greening Earth Society. This organization has published research reports with titles such as *In Defense of Carbon Dioxide* (New Hope Environmental Services n.d.), *Does Price Matter? The Importance of Cheap Electricity for the Economy* (Mills, McCarthy, and Associates 1995), and *The Internet Begins with Coal* (Mills 1999). This organization is funded predominantly by the Western Fuels Association. In the words of their website, "The Greening Earth Society is a not-for-profit grassroots organization created by Western Fuels Association to promote the viewpoint that humankind is a part of nature, rather than apart from nature."[3] Chris Paynter, the executive director of the Greening Earth Society, describes the Western Fuels Association. Western Fuel is "just coal; coal and electric utilities. They're a coal-supplied cooperative, supplying coal to consumer-owned cooperatives" (interview, 2000). In short, these organizations that call themselves "nonprofit organizations" have used money from for-profit industries to work against climate change regulations domestically as well as lobby on an international level to stop support for the Kyoto Protocol.

In 2000, the Global Climate Coalition changed its structure, becoming an association of trade associations that included the American Iron and Steel Institute, the American Petroleum Institute, and U.S. Chamber of Commerce. This new coalition "collectively represent[s] more than 6 million businesses, companies and corporations,"[4] and no longer allows individual company membership. It is unclear exactly what motivated the change in the membership structure. First, leaders of the coalition such as Glenn Kelly, the executive director and chief executive officer, say only that the coalition "restructured" (interview, 2000); however, leaders of environmental organizations such as Kert Davies—the science policy director of Ozone Action, which was absorbed by Greenpeace USA in 2000—claim that the reorganization was brought about by the fact that many of the major company members dropped out of the coalition. "Ford and . . . Chrysler just pulled out of this Global Climate Coalition . . . and did not renew their membership" (interview, Davies 2000). Whatever the motivation behind the change, the organization changed its position somewhat, recognizing that climate change is a real issue and emphasizing options that it considers "more constructive." In the words of Kelly, "If you want to step up to the table and talk about climate change, we have laid our solutions out" (interview, 2000). Although Kelly and his colleagues are prepared to discuss voluntary climate change mitigation, they have yet to turn their words into actions.

It would not be accurate, however, to conclude that U.S. industry has failed to demonstrate any of the kinds of leadership seen in Japanese and Dutch industries. For example, unlike the Global Climate Coalition, the Edison Electric Institute has coordinated programs that account for actual emission reductions. William Fang of the Edison Electric Institute, which represents the U.S. electricity companies that produce 70 percent of all electricity in the United States,[5] spoke about the interests of the companies his institute represents: "We are very much sincere and progressive in our attempts to do something about reducing greenhouse gases" (interview, 2000). Even Rafe Pomerance, the former U.S. deputy assistant secretary of state and climate change negotiator from 1993 to 1999, recognized industry's interest in the issue of climate change. "Plenty of them are saying it's a problem" (interview, 2000). In fact, Katherine Silverthorne, climate policy officer at the World Wildlife Fund (WWF),[6] stated during an interview at the COP-6 negotiations that, because of this type of industrial support for dealing with climate change, "businesses have changed, and industry is better than Congress now" (interview, 2000).

Beyond verbal claims, many companies have enrolled in voluntary projects to reduce greenhouse gas emissions. One of the largest and most well-known of these projects is the program sponsored by the Pew Center on Global Climate Change. The center was founded by Eileen Claussen, a former U.S. climate change negotiator. In her own words, "I put together a group of major U.S. corporations, which now number twenty-one, which will probably number twenty-five in a couple of months and then maybe fifty or seventy-five in a year, who take what I think is sort of a rational approach to [climate change] . . . [that] we have to start dealing with it now" (interview, 2000). The Pew Center's mission is to "advance the debate through credible analysis and cooperative approaches."[7] One of the center's main projects is to coordinate the Business Environmental Leadership Council to "make significant progress in addressing climate change and sustaining economic growth in the United States by adopting reasonable policies, programs and transition strategies."[8] Beyond organizations and industries facilitating private climate change projects, the U.S. government has managed a number of voluntary programs with industry, the most famous of which might be the Energy Star labeling program. A climate change official at the Department of State explained the priorities of the U.S. government: "the U.S. government wants to focus on the private sector" (interview, official B).

In some ways, these voluntary climate change mitigation strategies look a lot like what proponents of ecological modernization theory would predict. At the same time, however, they have had exceptionally limited impact so far. Although many voluntary programs exist in the United States, the problem with voluntary action is that, in the words of Pomerance, "there are hundreds of voluntary agreements between government and industry. . . . Sure, they have some effect on a small scale but . . . you've got 260 million consumers of carbon" in the United States (interview, 2000). In other words, these voluntary top-down measures do not go nearly far enough in dealing with the level of greenhouse gas emissions put out by the United States.

Still, although some industries have begun to take on voluntary measures to deal with climate change, many companies continue to support the conservative nonprofit organizations described here, promoting the belief that no action should be taken by the United States on climate change. Davies describes these industry-funded groups' work on climate change, pointing out some important reasons why these organizations are so successful at getting their message about the Kyoto Protocol heard on Capitol Hill: "These groups are very effective, and they're more present on the Hill. They've got more money than us, and their message is a much easier one. It's like: don't worry, be happy. You know it's much harder than: change your life, change your behavior, get smart, you know, don't be an idiot, don't drive a big car . . . don't pollute" (interview, 2000).

U.S. Energy Infrastructure

Before I conclude that the difference has to do entirely with the simplicity of the message from the conservative nonprofit organizations, it is worth considering the possibility that another key reason for their success has to do with the specific material characteristics of the United States itself. Gardiner introduces the overall problem, in his explanation of the unique U.S. position on climate change:

> I think, in the end, there's a large group of economic interests who are happy with the way things are today and would be perfectly happy if the world did not change. And in the end, if we're going to deal with climate change, we must change the way in which we produce and use energy, and there are powerful economic interests who . . . prefer the status quo and oppose change. . . . Underneath it all, what's really going on here is the debate about that set of politics, and we believe we have to change: we believe that the future is in clean energy and not dirty energy. . . . I think there is a lot of nice policy arguments about whether the developing countries should do more or less, but, in the end, I think you'll find that the primary difference between political leaders now who are in favor of moving forward versus those who are opposed to moving forward is sort of where they stand on this question of the status quo. (interview, 2000)

What is necessary is to go beyond the rhetoric of this statement and to look more closely at the material energy infrastructure of this country. One aspect of that infrastructure, of course, is its sheer size: the United States is the largest energy producer, consumer, and net importer in the world (EIA 2000, 1), but another aspect of the energy infrastructure is its specific character. In the words of Fang, "In the U.S., you've got plentiful forms that can be cheaply transported, and that's why renewables haven't come in very much in our industry or anywhere else. So I think those help account for some of the differing policy responses to those kinds of situations" (interview, 2000). In simpler terms, the specific material substances with which the United States fuels its economy are different from those in other countries, and these differences can help to explain the position of conservative nonprofit organizations in America.

Table 6.1, which presents the overall energy consumption and production of the United States, Japan, and the Netherlands, shows that the difference is not merely a matter of total energy supplies coming from domestic sources. As can be seen, the United States and the

Table 6.1. Total Energy Consumption and Production by Source in Each Nation-State, 1997 (1,000 metric tons of oil equivalent)

	Japan	*Netherlands*	*United States*
Coal			
Production	2,356	0	561,928
Consumption	86,532	9,228	513,299
Consumption (%)	*17.4*	*12.7*	*25.0*
Oil and liquid gas			
Production	795	3,040	396,622
Consumption	271,600	27,590	854,506
Consumption (%)	*54.7*	*37.9*	*41.7*
Natural gas			
Production	2,009	60,574	442,115
Consumption	54,948	35,323	507,971
Consumption (%)	*11.1*	*48.5*	*24.8*
Nuclear power			
Production	83,148	628	173,658
Consumption	83,148	628	173,658
Consumption (%)	*16.8*	*0.9*	*8.5*
Percentage indigenous			
Source (%)	*20.8*	*87.2*	*77.9*

Source: International Energy Agency (1999).

Netherlands produce most of the energy they consume, while Japan only produces about a fifth of its energy needs. Figure 6.1 provides a graphical depiction of the relationship between consumption and production in these countries.

Figure 6.2, by contrast, may help to provide greater insights into the reasons why the United States and the Netherlands would have such different positions regarding climate change; those reasons have to do with

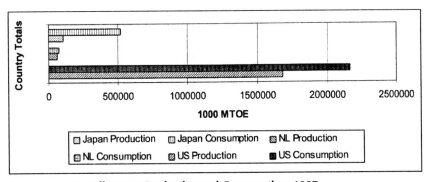

Figure 6.1. Overall Energy Production and Consumption, 1997
Source: Adapted from International Energy Agency (1999).

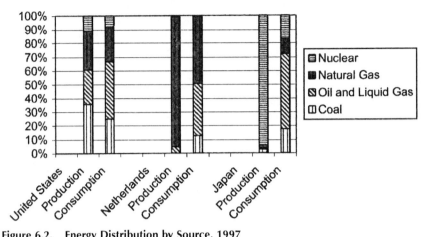

Figure 6.2.   Energy Distribution by Source, 1997
Source: International Energy Agency (1999).

the actual fuel sources being produced in each country. As this figure shows, although the Netherlands extracts natural gas from its reserves in the North Sea and although Japan produces nuclear energy, the United States produces high levels of oil, coal, and natural gas.

That the United States has become the most automobile-dependent country in the developed world is a fact related not only to the high levels of indigenous American oil but also to the open spaces within the country. As can be seen by the results of the quantitative analysis presented in chapter 3, dependence on motor vehicle travel is one of the best predictors of carbon dioxide emissions in the developed world. To stress these differences among the three case studies, table 6.2 lists the motor vehicle travel in the United States, Japan, and the Netherlands for 1997. In addition to the domestic oil reserves and the large open spaces, the high level of motor vehicle dependence in the United States is related to the national policy decisions regarding transportation infrastructure development, which (coming full circle) appear to be the product of the United States' large indigenous fuel supply (see Gramling 1996 for further discussion).

Perhaps even more significant than the high degree of American dependence on oil and automobiles is the country's dependence on coal.

Table 6.2.   Vehicle Usage in the United States, Japan, and the Netherlands

	United States	*Japan*	*Netherlands*
Motor vehicle travel 1997 (per billion kilometers)	4,090	756	109

Source: OECD (1999).

This point is particularly significant given the differences between the CO_2 emissions from the consumption of natural gas versus those from coal. "Coal releases more CO_2 per unit of generated energy than does oil, and oil more than natural gas" (International Energy Agency 2001, 27). As the *New York Times* reported, in July 2001, "American coal-powered plants pump 2.3 billion tons of CO_2 into the air each year—twice as much as the amount emitted by cars" (Goodell 2001, 6). Hugh Pitcher, staff scientist of the Global Climate Change Group at the Pacific Northwest National Laboratory, spoke about the implications of this difference: "You could roughly meet the Kyoto targets for the United States if you shut down every coal-fired electricity generating plant and replaced it with a combined cycle gas turbine" (interview, 2000). This statement highlights the fact that overall energy consumption and production are not the only variables that are important when considering the debates surrounding climate change policy making; rather, it is significant to remember the actual material that fuels the economy of each nation-state—in this case, coal.

In other words, as Pitcher's observation suggests, what is more important than the overall levels of energy consumption may be the fact that over 23.7 percent of the energy consumed is fueled by coal in the United States. Perhaps more important than the overall consumption of coal is the fact that the United States has the largest coal reserves in the world. Turning once again to table 6.1, we can see that the United States produces more coal than it consumes. The significance of this point is stressed by Davies: "Who will get hit [by the Kyoto Protocol] is coal, not so much oil. . . . If you're a power plant or a power company, you're in deep trouble, because they're the ones who really have to move. So I think the split has been more between, or is wider between, [first] these car companies and [second] the electric utilities [and] coal companies" (interview, 2000). This split that Davies refers to is significant in that it highlights the differences in the coal, oil, and automobile lobbies with regard to climate change. Although all three lobbies were members of the Global Climate Coalition in 1997, and although all worked to keep the United States from supporting the Kyoto Protocol, all of the U.S. automobile companies had dropped out of the coalition by 2000. In contrast, oil and coal companies continued to work through conservative nonprofit organizations to block American action to mitigate global climate change.

Both Davies and Pitcher speak of electricity production when they speak about coal consumption in America. Table 6.3 provides a list of the fossil-fuel share in electricity generation in the United States, Japan, and the Netherlands. As can be seen in the table, over 50 percent of U.S. electricity is generated from coal. In contrast to the American case, coal accounts for less than 30 percent of the electricity generated in the Netherlands and less than 20 percent in Japan.

Table 6.3. Percentage of Fossil Fuel Share in Electricity Generation in Select Countries, 1998

Country	Coal	Oil	Gas	Total
Japan	19.1	16.4	21.1	56.6%
Netherlands	29.9	3.9	57.0	90.8%
United States	52.7	3.9	14.7	71.3%

Source: International Energy Agency (2000c).

Given the high level of carbon dioxide emissions from coal, it is important to note why the United States continues to consume so much. In the view of Fang, if there were to be a transition to another fuel source, "tremendous amounts of natural gas would be needed; and the price, supply, the delivery of gas, those are all huge questions. And, of course, there are going to be other demands for the natural gas in other sectors of the economy" (interview, 2000). Fang, a lobbyist for the electricity industry, claims that an energy transition in the United States would be very expensive, if possible at all.

Whatever the actual expense involved in a transition to a different energy source, there is a clear question about the effect such a transition would have on the U.S. coal producers. As mentioned earlier in this chapter, the United States produces more coal than it consumes. Not only does the United States produce high levels of coal as a whole, but significant amounts of coal are extracted from twenty-six of the fifty U.S. states. In contrast to the U.S. oil reserves—which are "concentrated

░░░ Major Coal-Producing States (producing over 23 million short tons/year)

▨▨ Lesser Coal-Producing States (averages around 19 million short tons/year)

Figure 6.3. Coal Production in the United States (by State)
Source: www.eia.doe.gov/cneaf/coal/statepro/imagemap/usaimagemap.htm.

overwhelmingly (84%) in four states: Texas (25%) . . . Alaska (24%), California (21%), and Louisiana (14%)" (Energy Information Administration 2000, 2)—sixteen states in the United States are classified by the Energy Information Administration as "major coal-producing states," and ten more are classified as "lesser coal-producing states." These states constitute what Leggett calls the "problematic heartland of coal" in America (1999, 249). Figure 6.3 presents a map of coal-producing states in the United States. Once again it is important to highlight the economic aspects of this material good: although there are qualitative differences between the coal being mined in different regions of the United States, the extraction and consumption of the coal produced in all of these states fuel U.S. power plants and the U.S. economy—which may have something to do with the proclivities of the nation's elected senators and representatives as well.

The importance of coal as an energy source and an economic good, when seen from the perspective of a coal-producing state, is stressed by the legislative assistant and counsel to Senator Michael Enzi, a Republican senator from Wyoming—a state that happened to be responsible for 31 percent of the U.S. coal extraction in 1999. "We also have . . . strong concerns and interests in coal. . . . Part of the things that the Kyoto Protocol would do would be to take out our ability to produce and utilize that coal. That would be the end of the state economy. . . . Those elements fund a lot of the state, and we actually have seen many big benefits that have come from that" (interview, Scholes 2000). Not surprisingly, Senator Enzi has been one of a number of senators from coal states who have proposed legislation to limit the Kyoto Protocol.

More broadly, with coal being extracted from 26 U.S. states, 52 of the 100 U.S. senators come from states in which coal production contributes significantly to the state economy. Even more states run their electricity plants off of the cheap and indigenous U.S. coal supply. Although the relationships between the social actors involved in the decisions regarding the development of a climate change regime are important to understanding the complexity of the U.S. case, the U.S. energy industry, along with the conservative nonprofit organizations that they have funded to represent their political interests, contribute significantly to the U.S. climate change debate.

Though the deputy director of the White House Climate Change Task Force during the Clinton administration was indeed correct when he stated that the United States "must change the way in which we produce and use energy" to meet the emission reductions stipulated in the Kyoto Protocol (interview, Gardiner 2000), the lack of support for such a change in the U.S. energy infrastructure seems to be widespread. A transformation of the United States away from coal as the main source of electricity for the economy would not only affect the owners of a few wealthy companies but would also affect those working in energy extraction, as well as more than half of the electricity consumers in the country. In addition, given the vast quantities of coal that are shipped to power plants, plus America's heavy dependence on automobiles, an energy transition in the United States would likely affect the transportation sector as well. These material characteristics of the United States are important and should be kept in mind while following the roles that other social actors have played in the U.S. climate change debate.

Science

Given these material characteristics and the prominence of conservative nonprofit organizations, much of the global climate change debate in the United States has focused on the uncertainty of science. Perhaps Lee best explains the uncertainty argument in his work on science (1993, 159): "Uncertainty can itself become a source of power." In the case of global climate change in the United States, we are provided with a clear example of such a reframing. Anthony Socci, former associate director of the U.S. Global Change Research Program who is now a senior climate science advisor at the Environmental Protection Agency (EPA), points out, "If there was no money involved, I don't think there would be much contention one way or the other" (interview, 2000). Socci recognized the main reason that the science of climate change was challenged in the United States: the economic implications of regulation.

Although the Clinton administration expressed its support of the science of climate change, many members of the U.S. Congress showed uncertainty. In the words of a top staff member for a senator who is heavily involved in the climate change debate, "There is no question that there has been warming since the nineteenth century, but the problem is figuring out the reasons for the warming" (interview, Foreign Relations Counsel 2000). Similarly, Dallas Scholes, the legislative assistant to Senator Micheal Enzi of Wyoming, stated that Senator Enzi "does have some concerns about the science of it. Mainly at this point I guess the most outstanding question that is there is: is it man?" (interview, 2000). This questioning of the science extends to include questioning the data on which science and policy are based. During a discussion about the amount of CO_2 emissions emitted by the United States, Enzi's legislative assistant argued that they had "seen too many government reports that are . . . fundamentally flawed. . . . We're very skeptical about the work that's being produced by the EPA. . . . We think it is more politics than actual science in application" (interview, Scholes 2000). Once again, it is useful to mention that scientific uncertainty can equal power, and uncertainty about climate change, in particular, has been used by coal producers in Wyoming and elsewhere to gain power within the U.S. debate about the Kyoto Protocol.

This skepticism about the science of global climate change is even clearer in the opinions of the industry-funded nonprofit organizations. For example, the executive director of the Greening Earth Society, the organization sponsored by the Western Fuels Association, stated:

> [T]he global warming issue had been overstated and . . . the science behind global warming is not uncertain, it's just wrong and the apocalyptic warming is based on a lot of computer models, not reality. . . . There has been no steady rise or steady, real steady decrease, but overall it has cooled. . . . The warming that did occur, occurred before 1950, before most of the carbon dioxide from industrialization was put into the air so that warming had to be natural, and not man-made like the global warming proponents say." (interview, Paynter 2000)

This statement was based on reports commissioned by the organization itself—all of which had been funded, in some way or another, by the coal industry. The results of these studies are in direct contrast with the conclusions of the peer-reviewed Intergovernmental Panel on Climate Change (IPCC; see chapter 2), as well as the United States' own National Research Council (2001).

Anthony Socci, senior climate science advisor at the EPA, who served as a scientific advisor to Al Gore when he was in the Senate, discusses the differences in the science of the issue:

> I would say that the experts working on this issue on a regular basis, . . . [who] publish all the time in the usual peer-reviewed climate journals, would basically agree. . . . I mean who among this group of people are the real scientists here. . . . I mean those who are publishing frequently in well-known climate journals [with] peer review. . . . How much of that [industry-funded] information has been through that kind of filter, and who are those folks. . . . Often times you'll get a scientist who has been doing a lot of work [and] is recognized by the community of scientists as . . . an expert, and then you get someone over here who . . . doesn't have half of that or even a quarter or a third, or whatever, with sufficient ties to organizations . . . and the public reads this as a locked controversy, a controversy that's locked. (interview, 2000)

During the process of researching this book, the IPCC was preparing its third assessment. The final reports of the working groups of the IPCC were published in early 2001, and the summaries for policymakers that are derived from the reports of each working group were adopted later in the year (IPCC WGI 2001; IPCC WGII 2001; IPCC WGIII 2001). In response to the conclusions of this third assessment review, a number of U.S. Senate committees held hearings on the science and technology of climate change. Examples of such hearings include the Committee on Energy and Natural Resources, on June 28, 2001; the Committee on Commerce, Science, and Transportation, on May 1, 2001; and the Committee on Environment and Public Works, on May 2, 2001. Witnesses for these hearings included scientists from national labs and universities; industry representatives; and senators, including one of the lead sponsors of the Byrd-Hagel resolution, Chuck Hagel, from Nebraska. These hearings coincided with the Bush administration's supposed work to develop a climate change plan that would be an alternative to the Kyoto Protocol. Since the Bush administration rejected the protocol in March 2001, some claim that the science behind the policy became less of a central debating issue. One potential interpretation of the changes may be that, with President Bush stating that his administration would not consider the Kyoto Protocol, there no longer was a political threat that would motivate employing scientific uncertainty, as Lee described (1993); instead, the power had already shifted into the hands of the economic interests that emphasized the uncertainty debate during the Clinton administration. As can be seen by how the science has been challenged by industry and conservative nonprofit organizations, the precautionary principle (which is supported in Japan and the Netherlands) does not seem to be supported by the corporate sector or by national policy in the United States.

The State

As we turn now to the role that the state has played in the formation of the U.S. climate change regime, it is important to keep two points in mind. The first involves the significant influence that the energy industries and the conservative nonprofit organizations have had over many representatives in Congress and the Senate, as outlined earlier; the second has to do with the roles that the executive and legislative branches of the U.S. government play in ratifying an international treaty and bringing it to legal force.

As a number of my interviewees point out, the U.S. executive branch, led by the president and his administration, is responsible for negotiating a ratifiable treaty. The ratification itself, however, takes place after a final draft of the text is agreed on by all parties of the treaty. In the United States, ratification of the final text of a treaty takes place in the Senate. In other words, the president and his team negotiate the text of what they hope will become a treaty through the approval by the Senate.[9] Before any treaty can be ratified, however, the entire U.S. Congress must approve of implementing the legislation that will enable the United States to meet the requirements of the treaty. To clarify, before the United States Senate can approve the Kyoto Protocol, both sides of the Congress would have to agree on how the requirements put forth by the protocol would be met. This unique aspect of the U.S. case is pointed out by Pomerance, who states that this requirement is "not the same elsewhere necessarily and not widely understood. We cannot ratify without implementing legislation" (interview, 2000).[10] In other words, to ratify the Kyoto Protocol, Congress must pass laws that will result in the United States' meeting its emission reduction commitment of 7 percent below 1990 levels. Pomerance elaborates: "The issue is that you would have to have something that said that U.S. emissions were going to be minus seven [i.e., reduced by 7 percent below 1990 levels] . . . including whatever we buy abroad, but you're never going to get that through without knowing what the cost is" (interview, 2001). As will be spelled out in the rest of this chapter, these specific characteristics of the U.S. political process make consensus within the different branches of the government during the negotiation process very important.

As a result of the characteristics of the U.S. case described here, the history of the politics of climate change in the United States has long been one of debate and discord: since well before the Kyoto round of negotiations, in 1997, the United States had not had a consistent climate change policy, let alone one agreed on by the different branches of the government. The following sections briefly outline the steps leading up to the Kyoto round of the climate change negotiations.

UN Framework Convention on Climate Change

As was outlined in chapter 2, the United Nations Framework Convention on Climate Change (UNFCCC), written in 1992, was the first international climate change treaty. The Framework Convention only included targets for developed nations, and its goal was the stabilization of greenhouse gas emissions at the 1990 level. The goal was strictly voluntary, due in part to the fact that the elder Bush administration did not support binding emission reduction targets during the UNFCCC negotiations at the Earth Summit in 1992 (Leggett 1999). Even though the targets were nonbinding, the "Senate floor debate on ratification of the treaty brought out concerns by some Senators about the cost of compliance, its impact on the country's competitiveness, and the comprehensiveness with respect to the developing countries—concerns that were overcome because of the nonbinding nature of the reduction goals" (Parker and Blodgett 1999, 3). As early as the October 1992 Senate debates, the Democrat-led U.S. Senate—of which 52 of the 100 members come from coal-producing states—began to express apprehension regarding the international regulation of climate change. Even before the Clinton administration entered the White House and before the majority of the Congress shifted to the Republicans, the Senate expressed concerns about developing countries being exempt from any climate change treaty.

In response to the UNFCCC, the elder Bush administration and the Clinton administration each implemented a series of climate change action plans in 1992 and 1993 (U.S. Department of State 1992; Clinton and Gore 1993), with the aim of stabilizing emissions. Since the measures within the climate change action plans were, like the treaty itself, voluntary, U.S. greenhouse gas emissions continued to climb throughout the 1990s; in fact, prior to the negotiations in Kyoto, in 1997, U.S. carbon dioxide emissions were 12.9 percent above the 1990 levels (IEA 2001). In other words, even with climate change action plans, the United States had failed to meet its voluntary reduction targets. The Clinton administration blamed the failure to meet these reduction targets "primarily on unanticipated economic growth and on Congress not fully funding the programs" (Parker and Blodgett 1999, 10). The fact that the Democratic Congress of 1993 did not support the administration's budget to fund its climate change action plan may have provided the first sign of a disagreement between the administration and Congress regarding the issue of global climate change.

Also in 1993, while the Congress was still under a Democratic leadership, the recently elected Clinton administration tried, and failed, to implement a meaningful British thermal unit (BTU) tax. The proposed BTU tax was an energy tax that "would have taxed virtually all forms of fossil fuel energy in the U.S. and increased energy prices for end users (consumers and producers)" (Baron 1997, 9), but it never passed. The failure of this measure was in part due to the extensive reach of U.S. energy interests. High-ranking officials in the Clinton administration, such as Pomerance, believe that the failure of the BTU tax had a huge effect on climate change policy in America. In his own words, "The thing that drove us was the failure of the BTU tax. That was the limitation on domestic policy. Had the Clinton administration passed that, we would have been global leaders on the subject. . . . It was put up in part because of the climate issue, but it wasn't played as a climate measure; it was a budget deficit reduction measure. Had it passed, all would have been different" (interview, 2001; for more details regarding the BTU tax, see Baron 1997). Outside of the Clinton administration, members of Congress also saw this failure as a turning point. In the words of a senior staff member for the Democrats' side of the Senate Committee on Energy and Natural Resources:

I, like a lot of people, view that as one of the . . . mistakes that a lot of administrations make in the first year . . . coming out with something without . . . really working it. At the same time, the other constituencies that insist if you do not do it now [that] you have got political capital, you will never be able to do it. It was a disaster, it didn't happen. We wound up with a 4.3 cent increase to the gasoline tax, which has virtually no impact on consumer behavior, and we lost the majority in the Congress. I attribute a lot of that to that tax hike that first year, and it was passed on a partisan basis. (interview, senior staff 2001)

As this senior staff person states, the newly elected Clinton administration made mistakes with the BTU tax, and that failure had long-term effects.

Much of the discord surrounding U.S. climate change policy can be traced back to disagreements between the Clinton administration and representatives in the Democratic-led 103rd Congress that began after the UNFCCC was ratified, such as this debate surrounding the BTU tax. With the change in the majority of the Congress to the Republicans in 1994, these disagreements became more intense. It is important to note that this change took place during the early rounds of climate change negotiations, before the Kyoto Protocol was even written. The actions taken by both the Clinton administration and the Congress, after the failure of the BTU tax, resulted in a significant partisan rift between the executive and legislative branches of the United States government—a bad situation for attaining the overall goal of achieving greenhouse gas emission reductions, as well as for negotiating a ratifiable and implementable international treaty to mitigate global climate change.

COP-1

Prior to the first Conference of the Parties convening in Berlin in 1995, the U.S. Congress, which had already been hesitant to regulate energy consumption and climate change through a BTU tax, shifted to a Republican majority. Although the disagreements between Congress and the administration about regulating climate change can be seen as far back as the UNFCCC, in 1992, the rancor between the Clinton administration and the Newt Gingrich–controlled Congress regarding climate change was strongly shaped by this round of negotiations in 1995 and, specifically, to the resolution coming out of the first Conference of the Parties: the Berlin Mandate. Given the failure of most nations to meet their voluntary commitments to stabilize their greenhouse gas emissions, it was expected that the next international agreement would include legally binding and enforceable targets. Thus, the first round of negotiations focused on an agreement that stated that this next treaty would be consistent with the UNFCCC, with developed countries taking the lead in the first commitment period.

More specifically, the Berlin Mandate outlined "the rules of the road for the next phase of the negotiations" (Leggett 1999, 202). Most significant to the mandate is the fact that, like the UNFCCC, the treaty coming out of these negotiations would exclude developing countries from any commitments.

Although the mandate itself was agreed on by all the parties to the UN-FCCC, the U.S. delegation only signed the mandate after changing its position "at the eleventh hour" of the Conference of the Parties (Leggett 1999, 247). The U.S. delegation was aware that there was a good deal of concern in the Senate about the issue of developing country involvement. As an attempt to show its interest in continuing global progress on the issue, the U.S. negotiating team, representing the Clinton administration, ultimately signed the mandate.

In the words of a senior administration official in the Clinton White House who was not involved in the decision, "My sense was not that this was a position or result that was at all desired by us but that you couldn't get the deal done without it" (interview, senior official 2001). Similarly, Pershing recalls that the decision in "Berlin was a political call" that the Clinton administration signed because they thought of themselves as *the* environmental administration (interview, Pershing 2001). This decision to move against the Senate's interests opened a gulf between the Clinton administration and Congress that endured throughout the following years of negotiations. A member of the congressional staff for the Commerce Committee in the House of Representatives, who was a member of the congressional delegation in Berlin, noted that the administration's shift to supporting the Berlin Mandate was a critical point in the future of the politics of climate change in the United States. In this person's words, "We didn't think very much of Berlin when it came out. We said, wow, this is not going to lead to a good agreement" (interview, congressional staff 2000).

During my interviews, climate change leaders in the State Department also brought up the Clinton administration's signing of the Berlin Mandate, pointing out that their signing of the mandate has haunted them since 1995. As put by a senior administration official in the Clinton White House, "The Berlin Mandate has, in my view, been no end of grief. . . . It is not that the developing countries should be asked to do so much, but it is just that it is very difficult to conduct a negotiation when you start out saying that you do not have to do anything, and then you have to try to reel them back in. . . . The Berlin Mandate gets cited back at you three hundred times a day" (interview, senior official 2001). Other members of the administration also privately expressed their regret over the signing of the Berlin Mandate. During a meeting in 2000, for example, one State Department official argued that, although the mandate would limit greenhouse gases first in the developed world, the issue of climate change is a "global problem needing global solutions" (interview, official A 2000). In other words, despite the signing of the mandate, perhaps as an attempt to quell concerns in the Senate, the State Department continued to work to talk developing countries into playing a role in international climate change mitigation during the first commitment period.

COP-2

To be fair, although the Berlin Mandate excluded the participation of developing countries, it "had no legally binding requirements at all" (interview, Pomerance 2001). At the next round of the negotiations, in Geneva, the parties moved to agree to include legally binding reductions in greenhouse gases. Having gone against the wishes of the U.S. Congress during the Berlin negotiations, the Clinton administration was hesitant to proceed with signing on to this next step in the negotiation process. However, after holding back throughout much of the negotiations, the U.S. administration once again shifted away from the preferred position of Congress and supported legally binding targets that, because of the Berlin Mandate, would only apply to the developed world. In the process of changing its position, the United States broke from its alliance with Japan, Canada, Australia, and New Zealand—or the so-called JUSCANZ umbrella group. Leggett (1999, 249) describes them: JUSCANZ is a group of "developed countries with reasons to want to slow down the European Union. . . . The change in the U.S. position had cut the ground from beneath the JUSCANZ alliance." As a result of the United States' supporting the Geneva Declaration, the Kyoto Protocol, which was expected to come out of the COP-3 meeting the following year, would include significant legally binding emissions reductions for the developed world.

Senate Resolution 98

Because of its frustration with the Clinton administration, the Senate decided to make a clear statement about the parameters of the Kyoto Protocol that they considered acceptable. On July 25, 1997, over four months prior to the COP-3 negotiations, in Kyoto, Japan, the U.S. Senate unanimously voted to pass Senate Resolution 98, or, as it has come to be known, the Byrd-Hagel resolution. Highlights of the resolution include the following:

> Whereas the exemption for Developing Country Parties is inconsistent with the need for global action on climate change and is environmentally flawed; and Whereas the Senate strongly believes that the proposals under negotiation, because of the disparity of treatment between Annex I Parties and Developing Countries and the level of required emission reductions, could result in serious harm to the United States economy, including significant job loss, trade disadvantages, increased energy and consumer costs, or any combination thereof. . . . The United States should not be a signatory to any protocol . . . at negotiations in December 1997, or thereafter. (U.S. Senate, Congressional Record, June 12, 1997, report number 105-54)

As can be seen in this text, the Byrd-Hagel resolution directly responds to the Clinton administration's support of the Berlin Mandate and the

Geneva Declaration, making clear the Senate's position on emission reduction targets and doing so prior to the administration's sending its negotiating team to Kyoto in December. Chuck Hagel, one of the sponsors of the resolution, put the matter simply: "The Byrd-Hagel resolution is a complete rejection of the Berlin Mandate" (as quoted in Leggett 1999, 267).

The lead sponsors of the resolution were freshman Republican senator Chuck Hagel, from Nebraska; and Democratic senator Robert Byrd, from West Virginia. As pointed out by a senior staff member of the Senate Energy and Natural Resources Committee, Hagel was handpicked to stop the Kyoto Protocol by the Global Climate Coalition, one of the conservative industry-sponsored nonprofit organizations discussed earlier in this chapter. In this person's own words, "The Republicans were looking for somebody to take the lead on climate change and the campaign against [the Kyoto Protocol], and look who they got to do that. . . . I was at his very first fund-raiser when they announced [that Hagel] was their lead guy on climate change" (interview, senior staff 2001). Senator Byrd, in contrast, made statements on the Senate floor during the discussions regarding Resolution 98, asserting that

> I, for one, believe that there is sufficient evidence of, first, a probable trend toward increased warming of the Earth's surface resulting from human interference in natural climate patterns. I believe that a steady increase in accumulation of carbon dioxide and other greenhouse gases in the atmosphere is taking place. I believe that there is some relationship between the warming trend and such accumulations, enough to justify our taking some action and taking it now. (U.S. Senate, Congressional Record, July 25, 1997, S8116)

In 1997, Senator Byrd was one of the most senior Democrats in the Senate, but he also represented West Virginia. Even though he expressed some confidence in the importance of the issue of climate change during Senate discussions of Resolution 98, his state was responsible for 14 percent of U.S. coal production in 1998,[11] and some accounts report that Byrd himself recruited most of the support for the resolution. In the words of Pomerance, "Byrd went and lined everybody up. . . . [He] walked around with the resolution, [asking Senators to] sign his resolution" (interview, 2000).

There are numerous interpretations of the goal of the resolution. First, people within the Senate—Republicans and Democrats—had an interpretation similar to that of a staff member of a leading senator on the issue, who stated that any interpretation of the Byrd-Hagel resolution as a rejection of the climate change negotiations is "revisionist history . . . [Byrd-Hagel] is the Senate giving advice, not rejection" (interview, legislative assistant 2001). In Senator Byrd's own words—in fact, in his speech on the floor of the Senate—he describes the resolution: It "seeks to provide the Senate's views as to the global climate change negotiations now underway" (U.S. Senate, *Congressional Record,* July 25, 1997, S8116). However, many members of the Clinton administration's interagency team working on climate change saw the resolution as a "poison pill," meant to kill any possible U.S. participation in a global climate change regime. In the words of a senior White House official during the Clinton administration, the resolution "basically said: Do not sign this unless developing countries take on commitments . . . similar to what the developed countries were taking on. . . . The Berlin Mandate says they could not and they would not. So it was just like you had to do this, but that cannot happen, so I think it was designed as a poison pill" (interview, senior official 2001). Most members of the administration felt that the resolution was not giving advice; rather, they saw it as an attempt by strong forces in the Senate to kill the negotiations.

Although a senior staff member of the Senate Committee on Energy and Natural Resources reported that the administration worked with Senator Byrd on the text of the resolution and pushed Democrats to support it (interview, senior staff 2001), members of the Clinton negotiating team have an altogether different account. Pomerance recalls that the resolution was drafted by industry lobbyists. He reports meeting with Byrd and his staff, but he found that "we had no effect on the draft; we tried to get him to change it" (interview, 2001). Even though the administration was unable to change the text of the resolution, they decided not to fight it on Capitol Hill. The reason behind this decision is summarized by a senior White House official during the Clinton administration:

> We saw it as being very highly undesirable . . . but we were basically looking at a situation where . . . it was a very short period of time to the vote. . . . Our sense of it was that there was absolutely no chance to stop this thing. That this was completely, it was cooked, the train was barreling down the track, and if we had tried to do anything, we might have gotten . . . fifteen votes or something and that mounting an effort to stop it and getting flattened would have been worse in our judgment than just letting it happen. . . . Taking the line, that we are generally supportive of the spirit of the resolution. Yes indeed, we don't think anything should be done to hurt the economy and yes indeed, developing countries should play a part. . . . We just tried to basically make the best of it and didn't oppose it because it was futile at that point. In effect ninety-five to nothing seemed like a less bad outcome than like eighty-two to fourteen. (interview, senior official 2001)

Members of the administration's climate change team, both in the White House and in the State Department, were taken by surprise by the resolution proposed in the Republican-led Senate, and that fact may help to explain the success of the resolution. It is still unclear, however, if the Byrd-Hagel resolution was intended by its sponsors to be a "poison pill" or if it was merely to announce publicly what Congress considered to be the "appropriate U.S. position" (Parker and Blodgett 1999, 10) so that the Clinton administration would have a more difficult time negotiating on behalf of the United States without taking into consideration the Senate's position. If the actual intention of the resolution was to kill the Kyoto Protocol, then the issue of including developing countries may have been meant merely to stop any progress toward ratification. Given the unanimous vote, the resolution appears to have been more than just a partisan effort to block the political strength of the Clinton administration. At the same time, given the Republican-led Senate, President Clinton would not have been able to block its passage. Perhaps the fairest interpretation of the Byrd-Hagel resolution is that different members of the Senate supported it for altogether different reasons. It is quite probable that some senators, particularly Democrats, signed it because Byrd—a senior Democrat in the Senate—asked for their support; however, others strongly supported the message within the resolution: that a treaty that would follow the rules of the Berlin Mandate and the Geneva Declaration would be harmful to their state's interests.

Although in many cases congressional resolutions are nonbinding—frequently nothing but "acts of friendship or rewards or whatever" (interview, Pomerance 2001)—Senate Resolution 98 seems to be not such a benign policy recommendation. The Byrd-Hagel resolution still stands today as a reminder that the U.S. Senate did not support the rules for the Kyoto Protocol that was to be drafted at the COP-3 negotiations. Given their disapproval going into the drafting of the treaty, ratification and implementation of a revised text by the United States in the future seemed unlikely. That said, many agreed that, were there to have been another vote on the resolution in 2001, it would not have been unanimous. In the words of the staff member of a leading senator, the resolution "could never be done again," given that senators became more aware of the issue of climate change and that many were more likely to support the Kyoto Protocol (interview, legislative assistant 2001). Whether the resolution itself maintains the same level of support as it had in 1997, the consensus in the Senate continues to follow the opinion put forth by industrial lobbyists such as Edward Yawn, the director of government relations of the Edison Electric Institute, the association of U.S. shareholder-owned electric companies:[12] Senate Resolution 98 "is a key indicator of where the Senate is on the protocol and until the twin conditions [those of developing countries' involvement and a full outline of the economic implications of implementation] are satisfied, it is difficult to see how any administration would send the protocol up for advice and consent from the Senate" (interview, 2000).

COP-3

Even with this warning from the Senate about their support of a protocol that formed out of the rules developed in Berlin and Geneva, the Clinton administration sent a negotiating team to Kyoto and came back with the Kyoto Protocol—an international agreement that set legally binding emission targets for the developed world. To be fair, the administration had intended to return to the United States with a compromise between the United States' and the European Union's perspectives. As a State Department official and member of the U.S. negotiating team reported, the team went to Kyoto with the intention of agreeing on, at the most, a 1 to 2 percent emission reduction below 1990 levels. Pomerance, who was one of the negotiators in Kyoto, recalls that "the Dutch had gotten the EU to agree to negative 15 percent, and we allowed them to box us in, because we never said anything until the last minute. So the U.S. was boxed in by the time we got to Kyoto around the numbers. . . . We could have come out of there, I think, where we wanted, had we held firm" (interview, 2001).

Similar to its lack of organization in dealing with the introduction of the Byrd-Hagel resolution, the Clinton administration was not particularly organized in achieving its goals at the negotiations. It was only in June 1997, one month before the Byrd-Hagel resolution was voted on and less than six months before the COP-3 negotiations would be held in Kyoto, that the Clinton administration named a member of the White House staff to manage the communications and outreach of the climate change issue. Before then, in the words of a senior White House official in the Clinton administration, "The issue was not terribly high on the White House radar screen up until the G7 summit [happened in Colorado in June 1997]. Obviously there were a lot of people in the administration working on it, but it hadn't really cracked through the level where senior people in the White House were terribly focused" (interview, senior official 2001). It is likely that this lack of organization within the Clinton administration explains part of the failure of the administration to negotiate a politically viable treaty for the United States or to block the damaging resolution in the Senate.

The disagreement about the percentage of emission reductions continued at the Kyoto negotiations without any progress, and Vice President Al Gore, who had allotted a large part of his *Earth in the Balance* (1992) to the issue of global climate change, arrived in Kyoto to oversee the negotiating team. The team promptly shifted its position and agreed to a compromise: 7 percent emission reductions below 1990 levels. As Flavin reports (1998, 13), "The Clinton–Gore Administration could not afford to be found holding the noose if the Kyoto agreement was strangled." Particularly given

Gore's stated dedication to the environment, he was committed to return-
ing from Kyoto with a draft of a climate change treaty.

Supporting Flavin's interpretation is Eileen Claussen, the president of
the Pew Center on Global Climate Change, the former assistant secretary
of state for Oceans and International Environmental and Scientific Affairs,
and one of the lead U.S. climate change negotiators from 1993 to 1997. In
her explanation of the administration's actions at the COP-3 negotiations,
she notes, "They went to Kyoto because they wanted to make sure that it
was not a failure in Kyoto, that the U.S. would not be blamed for it. . . .
They agreed to something, and then they came back, and there was a
huge response from the people in the Congress and those in industry who
were totally opposed to anything" (interview, 2000). As Claussen points
out, members of Congress, many of whom attended the negotiations in
Kyoto, were not happy with what transpired at COP-3. For example, consider Scholes's summary of Gore's involvement: "When Vice President
Gore showed up and saw what had happened, he basically caved in and
gave up all of the U.S. concessions and gave up everything. . . . He
changed the policy and implemented his policy" (interview, 2000).

That this statement is made by a staff member of a senator from the
top coal-producing state in America reflects what may be the most extreme response to the outcome of the COP-3 negotiations. But perceptions in the Senate that Gore "caved in" in his agreeing to a 7 percent
emission reduction mirror accounts that are presented elsewhere (see,
e.g., Leggett 1999). Still, members of the Clinton administration had an
entirely different interpretation. A senior White House official reports
that "it certainly would not be fair to say that Gore came in and threw
his own pitch and nobody knew he was going to do it. . . . He didn't hi-
jack anything. . . . We were in quite close consultation with the White
House back home, and we were calling all the time" (interview, 2001). In
addition, the administration defended its decision to agree to larger
emission cuts by stating that, through its accounting, the 7 percent emission reduction would have been only a 3 percent reduction (interview,
senior official 2001). In fairness, the fact that many accounts of the negotiations state that Gore "betrayed" his country's interests in Kyoto
may well be reflective of the level of criticism for the Clinton administration's climate change policy by December 1997.

In the minds of many members of the Clinton administration as well
as the negotiators who represented them, the Kyoto Protocol, as it was
drafted in 1997, was only a work in progress. Indeed, the purpose of the
subsequent Conferences of the Parties was to finalize the text and
the mechanisms of the protocol so that it would be in a ratifiable form
for the parties. Although the Clinton administration had focused on the
protocol as a work in progress, others saw it as the final straw that

destroyed the last bit of confidence that the Senate and climate change opposition had in the administration's ability to negotiate for the United States. The negotiating team, led by Gore, did exactly what the Byrd-Hagel resolution cautioned against: they made the United States a signatory to the Kyoto Protocol. In the words of a top staff member for a senator who had been heavily involved in the climate change debate, "The Kyoto Protocol is never going anywhere in the United States because the administration and the negotiating team knew it wasn't going anywhere and still went forward with the politics" (interview, Foreign Relations Counsel 2000). At a minimum, this quote illustrates the significant lack of support within the Republican-led Senate for the Clinton administration during this period. To be sure, although parts of the protocol were still negotiable at the subsequent international meetings, neither the participation of developing countries nor the levels of legally binding emission reductions were still on the table for discussion; thus, whatever the Clinton administration may have claimed, the United States did not appear to be in a position to sign anything.

Post-COP3

After COP-3, the status of the Kyoto Protocol in the United States in particular, and that of the U.S. climate change debate in general, did not improve. The Clinton administration did focus on responding to the Senate's concerns, as can be seen in the words of David Doniger, director of climate change policy of the Environmental Protection Agency and a lead member of the U.S. negotiating team during the Clinton administration prior to the negotiations in The Hague, in 2000: "I'm hoping that we can bring back something [from COP-6] that would answer the concerns that were implicit . . . in Byrd-Hagel about economic impact on the country that we'll be able to assure members of the, well, first members of the public and stakeholders, that this is not going to have an adverse impact on them economically" (interview, April 2000). The Clinton administration continued to negotiate the Kyoto Protocol with the hope that they could negotiate a final draft that would meet some of the requirements of Senate Resolution 98. However, most of the requests for the inclusion of developing countries' voluntary engagement were refused, since these countries had become exempt from the Kyoto Protocol under the provisions of the Berlin Mandate, in 1995.

While the administration focused on finalizing the text of the protocol, Congress dug in its heels, with members proposing legislation to limit the administration's ability to implement aspects of the protocol, or to even spend significant money on studying the issue. One Senate staff member described the depth of the rift between the Congress and the Clinton administration, saying that the Kyoto Protocol set back the issue of climate change because it created partisan divisions, meaning that "now, even sensible things are tough that would otherwise be consensus policies" (interview, Foreign Relations Counsel 2000).

The opinion that no climate change policy was possible during this post-COP-3 stalemate was mirrored by members of the administration. Gardiner called the situation "a real political brawl" (interview, 2000), and a senior staff member of the Senate Committee on Energy and Natural Resources similarly described the politics during this era:

> In 1995 the Republicans took the majority in the Congress, and you know, as everyone knows, the Contract with America, and so . . . the Republican Revolution . . . blew into town and a lot of people, like the fossil energy industry, thought they would have a lot of support from the Congress to oppose what they perceived to be an overly green Clinton administration. . . . The problem for Democrats last Congress is that we started the Congress [in 1998—after the negotiation of the Kyoto Protocol] with the impeachment of Bill Clinton. By that time, the Republicans hated him with such a vengeance. The politics in Washington were just bitter, and bitterly partisan on everything, even things that you know had never been particularly partisan before. (interview, senior staff 2001)

This person's description of the increasing level of partisanship and congressional anger toward the Clinton administration provides some context in which to understand the difficult relationship, or what I call a "rift," between these two branches of the national government during this era.

The ramifications of this rift continued to haunt domestic actions on global climate change throughout the Clinton administration. Members of the House of Representatives, for example, led by Congressman Knollenberg, of Michigan, sponsored a successful rider to a number of different appropriation bills. Generally referred to as the "Knollenberg rider," it made it illegal for government agencies and federally funded organizations to spend their time working on Kyoto Protocol–related work. This rider, which had been applied to appropriation bills that determined aspects of the national budgets from 1998 through 2000, highlights the continued animosity in the U.S. Congress toward the Clinton administration's work on the Kyoto Protocol. Although many of the Democrats in the Republican-led Congress would have supported climate-change-mitigating policies by 1998, Congress continued to halt any progress on climate change with the help of Democrats such as Knollenberg, who is from the district that houses the industrial headquarters of the Ford Motor Company—Detroit.

Pomerance, however, interprets the appropriations rider differently, saying that it "was bullshit." According to Pomerance, Knollenberg said that the rider was to protect the United States against "back-door implementation" of the Kyoto Protocol. But, in Pomerance's words, "That was rhetoric. The fact is that the only way to implement [the protocol] was implementing legislation. The argument [implicit to the rider] was that we should do nothing on climate change because it is all an attempt to implement Kyoto . . . but we are under obligations under the Framework Convention. . . . We could do stuff under that" (interview, 2001). In sum, this appropriations rider was just another example of the level of animosity between the Clinton administration and members of the Congress.

Whatever the true motivation behind the rider, Nathalie Eddy—the international coordinator of the U.S. Climate Action Network, a network of proenvironmental nongovernmental organizations working on the issue of global climate change—discussed the implications of the rider: "All the agencies have their hands tied, and they're trying to dance around the language so that they're able to do their jobs without mentioning Kyoto or climate change" (interview, 2000). As a result of policies such as this rider, the Clinton administration avoided policies that required the support of the Congress for implementation. Claussen perhaps best summarizes the implications of the Clinton administration's policies since the drafting of the Kyoto Protocol:

> [The administration has] done a few things that are relatively easy, like executive orders for how the government should run its own shop. . . . They've been frightened off by some sort of vigorous and vociferous people in the Congress who want to make sure that there is no "back-door implementation of Kyoto," and the result of that is that there has not been any reasonable dialogue on Kyoto. . . . So the United States has continued to work abroad to better define what was in Kyoto . . . but they haven't had a serious, in my view, a serious domestic effort, which makes the international hand much harder to play as well, because if you aren't doing anything at home, it's sort of hard to be such a big bully overseas. (interview, 2000)

As Claussen points out, the legislation passed in the Republican-led Congress meant that the Clinton administration was limited in its policy-making ability, and the lack of climate change policies coming out of the United States in the years following COP-3 had weakened the position of the United States in the international arena. Basically, no real climate change policy was possible because of, first, the administration's disorganization in the early stages of the international policy-making process; second, the extreme polarization between Congress and the Clinton administration; and third, the efforts by conservative nonprofit organizations representing the interests of the energy industries. In the view of Pitcher, no real political action could happen in the United States during the end of the Clinton administration "given the current polarization" (interview, 2000).

With the lack of political action regarding the issue of climate change, most people expected a resolution after the 2000 election. Almost every person whom I interviewed prior to the 2000 presidential election pointed out that the future of the Kyoto Protocol in the United States would be contingent on the outcome of the election. Yawn, however, argues that the resolution of the rift between the administration and Congress would be dependent on more that just the presidential elections: "Even if Gore won the election and the House changed over to the Democrats, you still have a Republican Senate and, as long as you do, you know the resolution, Senate Resolution 98, stands" (interview, 2000). Yawn, an industry representative, acknowledges that, although the Byrd-Hagel resolution was passed unanimously, the strongest supporters of the resolution in the Senate were Republicans; only with a Democrat-led Senate might the resolution no longer hold.

Issues related to the presidential election resonated throughout the sixth Conference of the Parties. Because of the timing of the negotiations and because the negotiators were appointed by President Clinton, the negotiating team at The Hague was that of a "lame duck administration" (interview, Claussen 2000). A staff member of the House Commerce Committee pointed out the difficulties of the timing of the COP-6 negotiations. "It's going to be especially tough because you're going to have two weeks into a new administration with no one on either side having a real mandate on the U.S. side to agree to anything" (interview, congressional staff, August 2000). This situation was made even more difficult due to the election problems in the United States; no one at the COP-6 meetings knew if Gore or Bush would be the next U.S. president, not even the U.S. negotiating team. It is probable that this lack of certainty regarding the future U.S. leadership played a role in the failure of the negotiations, which were later set to resume in July 2001 in Bonn under the title of COP-6bis.

The election crisis resolved itself, and people involved with the international climate change negotiations expected the U.S. position to reset, with the new administration supporting the regulation of carbon dioxide emissions and the Kyoto Protocol. After George W. Bush took the helm in the United States, however, his administration was slow to present a clear position on climate change. First, the Bush administration reversed its position on the regulation of carbon dioxide as an air pollutant. Then, on March 13, 2001, President Bush responded to a letter sent to him from Republican senators Hagel, Helms, Craig, and Roberts, stating that his position was consistent with Senate Resolution 98. In his response to the senators, the new President Bush said that he would not consider a protocol that "exempts 80 percent of the world, including major population centers such as China and India, from compliance, and would cause serious harm to the U.S. economy."[13] After many years of disagreement involving the administration and

Congress, the newly elected President Bush finally closed the gap between the positions of the Republican-led Senate and the administration.

Bush's response to the issue of the Kyoto Protocol temporarily unified the U.S. government's position on the issue of climate change—albeit a position different from any other country in the world. Since Bush's statement, however, the administration repeatedly delayed releasing an alternative climate change policy, which was promised in March 2001. If the president and the Republican-led Senate had been merely critical of the protocol itself as a tool for mitigating climate change but were sincere in their desire to do something about the issue, the Bush administration would have offered an alternative proposal, which it promised after rejecting the Kyoto Protocol but did not release until early 2002. The fact that the Bush administration offered an alternative proposal for climate change mitigation so late suggests that Bush's intention was not to respond to the Senate's concerns raised within the Byrd-Hagel resolution but, rather, to kill any possible regulation of climate change in the United States. To understand Bush's position on climate change, it is perhaps useful to note that, as was reported by the *New York Times*, Bush comes from "the state that consumes more coal than any other in the country, together with [Vice President] Dick Cheney, who hails from the largest coal-producing state in the country" (Goodell 2001, 6). Once again, we are reminded of the material characteristics of the United States when we try to understand the U.S. position on climate change.

Even though the Bush administration reversed its position on climate change to reflect that of the energy industry and the Republican leadership in the Senate, the leadership of the Senate changed to the Democrats when Senator Jeffords left the Republican Party in May 2001. With the Senate being led by the opposite party than the one that governs the White House, the consistency of the climate change policy in the United States was once again uncertain. In fact, on June 8, 2001, Senator Byrd— one of the lead sponsors of the Senate Resolution 98—proposed the Climate Change Strategy and Technology Innovation Act of 2001. The goal of this bill was to begin to address the issue of climate change within the United States. Given the fact that Byrd comes from the third-highest coal-producing state in the United States, it is hard to tell if his activities represent an actual interest in climate change mitigation or if they are examples of more symbolic types of action, which have been addressed by Edelman (1964) and Buttel (2000a). Given Bush's lack of interest in reengaging with the global community on the issue of climate change, the success of Byrd's bill is no threat to the U.S. position, as the United States will not be a party to the Kyoto Protocol. The legislative assistant for a senator involved in the climate change debate pointed out that the lack of a U.S. climate change policy leading into the COP-6bis round of negotiations in

Bonn would affect Senate attendance at the negotiations: "This could be an uncertain and possibly difficult negotiation. I would doubt any members will go, and staff would be limited" (legislative assistant, correspondence with author, June 2001).

Civil Society

It was this battle between the Democratic Clinton administration and the Republican-led Congress, along with the efforts of the conservative non-profit organizations representing American energy industries, that resulted in a discourse of uncertainty regarding the issue of climate change in America. Given this situation, it is not a surprise that U.S. civil society has been relatively inactive in its support for the Kyoto Protocol. As has been outlined in this book, many types of civic engagement are bulked under the term *civil society*; ranging from individual citizen action to citizen action through professional social movement organizations (SMOs). Unlike the Dutch case, where SMOs work with the government and citizens work through consumer-based activities for the same goal, many SMOs in the United States tend to work outside of the political process, but they also appear to receive little support from the public at large. This difference is highlighted by a senior staff member in the Senate Committee on Energy and Natural Resources, who identifies environmental SMOs as "rabid enviros" that were responsible for the failure of Democrat-led efforts in support of early action against climate change. In the staff member's own words, it "didn't fall apart because of the industry. [Support for the bill] fell apart because of the enviros. They savaged them for it. Ozone Action, Sierra Club, World Wildlife Fund, a bunch of the extreme enviros just went after [the senators] on it and made it impossible for a lot of liberal Democrats to support it" (interview, senior staff 2001). Here, the staff member describes one attempt by Democrats in the Senate to pass a moderate first step in climate change legislation—perhaps all that was possible in the highly partisan Congress—but environmental groups were so critical of the effort that many Democrats felt pressure to pull out of the bill. Although the staff member was perhaps a bit harsh in characterizing these SMOs as "rabid enviros," members of the Clinton administration argued that environmental groups were not in line with the U.S. policy opportunities or with the American citizenry. In fact, Pomerance contends that environmental groups' intolerance of compromise made it very difficult for the Clinton administration to support the final offer proposed by the EU during the first part of the COP-6 negotiations in November 2000: "The Clinton administration's job was to come up with something that would work, not something that would please enviros, and I think we did too much of the latter and too little of the former" (interview, 2001). This claim about the failure of environmental groups in the United States to support proenvironmental efforts is consistent with the role that many American environmental groups played at the climate change negotiations. Instead of working with the U.S. government to create the most environmentally sound policy that was possible in the highly partisan political environment during this time in the United States, these environmental groups are generally seen as having led much of the anti-U.S. criticism at the negotiations with groups from the EU and elsewhere. Given the extent to which the environmental groups had been excluded by the economically conservative debate taking place in the United States, it is possible that they saw no role for themselves, except as critics outside the policy-making process. As Pomerance points out, the Clinton administration could have used the support of SMOs for its more environmentally sound policies when they were challenged by the Republicans in Congress.

Although environmental organizations continued to push the U.S. government about climate change, citizens did not seem to follow suit. With the Byrd-Hagel resolution defining the Senate's position since 1997, members of the Congress, especially those from the Republican majority, had "to hear . . . from constituents, and I don't think there's been much of that, if any" (interview, Claussen 2000). Gardiner reflected a similar opinion: "The public is the ultimate decision maker . . . about whether they believe strongly enough in the issue and feel we need to do something about it— that they basically want to let their elected officials know that and hold them accountable" (interview, 2000). Similarly, Pomerance argues that if there were "sufficient political demand," members of the Congress would have listened (interview, 2000). Although there may have been insignificant public pressure, it is not a surprise that the American public was confused and not particularly vociferous on the issue, given the high level of conflicting information being released into the public sphere about the issue of climate change from the conservative nonprofit organizations, the energy industries, the Republicans in Congress, the scientists affiliated with the IPCC's peer-reviewed process, and the Clinton administration.

Another potential explanation for why citizens did not make their voices heard involves the complexity of the issue of climate change itself. Even without all of the conflicting accounts of the issue, the phenomenon is not simple. Again, the construction of the science of climate change—in this case, in the United States—presents a case of Lee's notion that uncertainty can create power (1993, 159). This power can be seen in the way the issue of climate change in the United States had been reframed to be a debate about science; this reframing affected civil society, in making the topic even more difficult for laypeople to understand. In the words of a leading State Department official, "The American public is not very engaged in this issue. . . . It's because of the complexity of the issue" (interview, official A 2000). Although part of the reason that the American public was uninvolved in the issue was its overall complexity—which also limited many Japanese and Dutch citizens from getting involved—I contend that the inundation of competing interpretations of the science kept many Americans from mobilizing.

In polls that specifically ask about climate change, the majority of Americans are found to think that "global warming is a real problem that requires action" (Kull 2000, 2; see also "Time/CNN Poll: We Want Action" 2001). Still, a Gallup poll conducted right after Bush changed his position on the issue found that even if Americans do consider climate change to be a real problem and even if they do "believe human-induced increases in the Earth's temperature will occur in their lifetime, they do not believe the effects will be so severe as to negatively impact mankind" (Gallup Organization 2001, 3). Although U.S. citizens have been very active and effective in supporting the passage of some of the strictest environmental regulations in the world regarding air and water standards, citizens were relatively inactive on climate change due, in part, to the confusion surrounding the science of the issue of climate change—which, to state once again, is the product of different interest groups promoting the uncertainty of the science.

One final possible explanation of civil society's lack of involvement may be that Congress, by listening to the energy interests, had been already working on behalf of at least some citizens' interests. Although citizens consider climate change an issue worthy of "action," they may not be interested in the kinds of action that are included in the Kyoto Protocol. In response to a question regarding American civil society's interest in the climate change debate, a high-ranking Senate staff member brought up the gasoline price demonstrations that took place in the summer of 2000. Given that the citizens went "bonkers . . . it is incomprehensible that the Kyoto Protocol is possible here" (interview, Foreign Relations Counsel 2000). Similarly, a senior staff member in the Senate Committee on Energy and Natural Resources also spoke about citizen responses to the increased gas prices, pointing out that American fuel consumption continued to rise with the increase in oil prices. "People did not reduce their driving. . . . I mean, people bitch, and we get all these letters, and there's this whole uproar about it, but we'll see what actually happens in the statistics. . . . [If people really cared about climate change], you would see changes in habits, in behavior, in purchasing decisions. I mean . . . half of the vehicles sold in this country for the last four or five years have been SUVs" (interview, senior staff 2001). In fact, in a number of polls in the United States, the data mirror the opinions put forth by these congressional staff members: U.S. citizens are not willing to pay significantly to mitigate global climate change (see, e.g., Kull 2000; "Time/CNN Poll: We Want Action" 2001; see also, Gallup Organization 2001). That said, citizens' changing their lifestyles has been a particularly ineffective way of dealing with the issue of climate change.

The purchasing behavior of Americans, like their understanding of the issue of climate change, is the product of the lack of consensus regarding the issue put forth in the public sphere. Similar to the situation in Japan, citizens in the United States have high consumption levels. At the same time, the Japanese government is working with industry to ensure that energy-efficient options are available. With the large contingent of energy companies blocking any such measures in the United States, American citizens have not been given much opportunity to do more. Because of the stalemate that exists in the U.S. government regarding the issue, pressure from civil society seems to be one of the best opportunities to affect results in the climate change status quo in the United States.

CONCLUSION

This summary of the major actors involved in the U.S. climate change debate brings us to the Bush administration's decision not to consider the Kyoto Protocol. After his decision was made, there was significant international criticism of the United States, as well as from citizens within America. This attention to the U.S. position on the Kyoto Protocol contributed to criticism from Democrats in the Senate, as did Bush's budget for 2002, which included extensive cuts to a number of environmental programs and increased natural resource extraction. A senior staff member in the Senate Committee on Energy and Natural Resources for the Democrats reported that "at some point, we'll start moving forward on a domestic agenda" (interview, senior staff 2001). Similarly, Bryan Hannegan, the staff scientist for the Republican side of the same committee, outlined possible actions coming from the Republican side: "We have sort of laid the groundwork for several legislative ideas that will be before the Senate" (interview, 2001).

After the majority in the Senate changed to the Democrats, hearings were held on the Climate Change Strategy and Technology Innovation Act of 2001, which was proposed by one of the lead sponsors of the Byrd-Hagel resolution: Senator Byrd, of West Virginia. Although these discussions opened the door to domestic climate change action, it could only go so far toward reengagement in the international discussions: It is the administration that is responsible for negotiating any treaty, and the Bush administration made itself clear about its position. As mentioned, the Bush administration did not release the alternative to the Kyoto

Protocol until early 2002, though it was promised in March 2001. Similar to the lack of communication between the Clinton administration and the Republican-led Congress, Hannegan reports comparable communication problems between the Republicans in Congress and the Bush administration, during the period when the Republicans held the Senate majority:

> There is not that much dialogue between the branches on this. I think it stems from the notion that the president goes out, or the agencies go out, they negotiate these things, and they drop them before the Senate, and the Senate is sort of the arbiter of whether [the protocol] is good or not for us. There is not a whole lot of give and take, and even now with the new administration, you know, you have got this climate review going on and it is very closed off and a lot of us have sort of wrung our hands collectively. (interview, 2001)

With a change to a Democratic-led Senate, the communication problems were likely to increase.

The global implications of these changes are even harder to predict. Even before President Bush took office, Tsuneyuki Morita stated that "the U.S. is failing on climate change" (interview, 1999; recall from chapter 4 that Morita is the head of the social and environmental systems division of the Japanese National Institute of Environmental Studies and a coordinating lead author of the IPCC's third assessment report for Working Group III, on climate change mitigation). Leading up to the COP-6bis meeting, in July 2001, there were different and conflicting reports about what other states would do about the Kyoto Protocol if the United States continued to refuse to be a party to the treaty.

Regardless of what happens internationally, it is necessary to understand the material reality of energy consumption and production in the United States, in particular that of coal, in combination with the conservative nonprofit organizations that are representing energy and other economic interests. These characteristics, in addition to the partisan rift between the poorly organized Clinton administration and the Republican-led Congress, contributed to the United States' climate change policy being one of debate and discord. On a theoretical level, whatever becomes of the U.S. climate change regime, one thing is certain: the United States does not represent a case of ecological modernization, reflexive modernization, or postmaterialism. Rather, it is possible that the issue of global climate change in the United States presents a case of materialism, with economic interests beating out all other sociocultural actors. To date, material characteristics have had the most powerful effects on the political outcomes of the United States' climate change regime formation.

The U.S. rejection of the Kyoto Protocol, in the face of international pressure, seems to go against the notion of the global environmental system at first glance; however, upon deeper consideration, it is consistent. As may be recalled from chapter 1, the global environmental system involves the idea that all countries are part of an entire system; these countries interact with one another. Since the end of the Cold War, no state has emerged that can compete with the United States. Thus, within this global system, the United States is in the privileged position of being the solitary economic and political superpower that can unilaterally pull out of an international treaty, and as a result of the interrelations between domestic actors described above, that is exactly what it did.

NOTES

1. See www.globalclimate.org/BushLetter.htm (accessed October 2, 2003).

2. See, for example, Harris (1998); Leggett (1999); Paterson (1996).

3. See www.greeningearthsociety.org/about.html (accessed October 2, 2003).

4. See www.globalclimate.org/aboutus.htm (accessed October 2, 2003).

5. Available at www.eei.org/about_eei/index.htm (accessed October 2, 2003).

6. Although some branches of this organization call themselves the World Wildlife Fund and others the World Wide Fund for Nature, they are all members of the international network and go by the acronym WWF.

7. See www.pewclimate.org (accessed October 2, 2003).

8. See www.pewclimate.org/companies_leading_the_way_belc/ (accessed October 2, 2003).

9. As stated in the U.S. Constitution: The president "shall have power, by and with the advice and consent of the Senate to make treaties, provided two-thirds of the Senators present concur" (article 2, section 2 [2]).

10. Although there has been very little written about this characteristic of the U.S. ratification process, Pomerance points out that, because of this requirement, the United States has yet to ratify the Basel Convention on hazardous waste.

11. See www.eia.doe.gov/emeu/cabs/usa.html (accessed October 2, 2003).

12. Adapted from www.eei.org/about_eei/index.htm (accessed October 2, 2003).

13. See www.globalclimate.org/BushLetter.htm (accessed October 2, 2003).

7

Conclusion

We now return to the outcome of the second part of COP-6—the COP-6bis—in Bonn in July 2001. As discussed in chapter 1, all advanced countries except the United States stated that they would continue to move forward on an international regime that is designed to begin to mitigate global warming. In fact, during the following fall, at the Conference of the Parties-7 (COP-7), in Marrakech, Morocco, the "Parties agreed [to] a legally binding compliance regime" (Vrolijk 2002, 4). Through the agreement coming out of this meeting—the Marrakech Accords—the domestic ratification process was finally ready to begin.[1] Since the meeting in Morocco, 119 countries have ratified the Kyoto Protocol.[2] Among them are Japan and the Netherlands, the latter of which provided one of the votes to allow the European Union's ratification of the protocol on May 31, 2002. Together, the ratified countries' emissions in 1990 represent 44.2 percent of the carbon dioxide emitted by the Annex I countries. Thus, the disparate responses to the Kyoto Protocol that have been discussed in the case study chapters of this book continue—neither position has won out completely, but the global climate change regime continues to develop, albeit without involving all of the nations for which it was intended.

LOOKING BACK AT THIS STUDY

The continuing events in the development of a legally binding climate change treaty provide more data regarding the central theoretical and empirical question of this book, as first spelled out in chapter 1: How are national responses to international environmental policies and their effects on the international policy-making process to be explained? As will be recalled, I turned to one particularly prominent issue—global climate change—in the effort to answer this broad question. Quantitative analysis of characteristics of the member nations of the Organisation for Economic Co-operation and Development show that, although environmental protection is possible through measures of ecological efficiency, the variations across industrialized nations—rather than the commonalities across those nations—provide the lion's share of explained variance in CO_2 emission levels. Thus, I selected three countries that represent particularly important cases in the global climate change debate, to examine more closely the dissimilar domestic responses of these nation-states to the potential regulation of an environmental good through an international treaty: the Kyoto Protocol. Although the differences among the responses to the protocol were scarcely visible on the international level in 1997, when the original text of the treaty was drafted, those differences between the responses had become very clear by the end of the 2001 COP-6bis negotiations, in Bonn. The EU pushed forward to implement the regime quickly; Japan continued to be dragged along, with other parties making concessions to keep it involved; and the United States rejected the protocol without any hint that it would reengage in the foreseeable future. In other words, since the meeting in Bonn, domestic ratification of the Kyoto Protocol has been consistent with these positions.

Unfortunately, existing theories of society–environment relationships prove inadequate for explaining what has happened to the Kyoto Protocol since it was agreed upon by states in 1997; neither the economic arguments nor the environmental arguments put forth by scholars of society–environment relationships win out completely. Instead, an adequate understanding of the present evolution in the global climate change regime demands a full consideration of the roles of science, industry, civil society, the state, and international actors.

My analysis suggests that, although what I have classified as the environmental state and environmental sociology theories fit specific aspects of the domestic climate change regimes presented in this book, neither fully explains the complex interrelations among actors on the domestic and international levels. A broader orientation is required. The concept of the *global environmental system*, which involves the interactions between domestic and international social actors, provides a more accurate model

Table 7.1. Climate Change Regime Matrix (updated)

Country	IV1 State	IV2 Science	IV3 Market	IV4 Civil Society	DV1 Political Outcome	DV2 Material Outcome[a]
Japan	Strong	Central	Collaborative	External-local	Ratified June 4, 2002	Increase 7.6% since 1997
Netherlands	Medium	Middle	Autonomous	Internal	Ratified April 12, 2002	Increase 9.3% since 1997
United States	Weak	Peripheral	Autonomous	External-national	Rejected May 13, 2001	Increase 11.7% since 1997

[a]Measures of emission reductions are calculated from the most recent CO_2 emission inventories (1998), which are published by the International Energy Agency (2001).

of the formation of national positions regarding the global climate change regime and the Kyoto Protocol. In sum, it is the interaction among the social actors in each nation-state that explains the variable political responses to international environmental policy making. This research provides both empirical elaboration and theoretical development of this framework, specifying the ways in which the relationships between national and global actors are set within the context of the continual interaction of science, government, corporate, and citizen forces.

DOMESTIC ACTORS WITHIN THE GLOBAL ENVIRONMENTAL SYSTEM

Within the case studies of the United States, Japan, and the Netherlands, there were many similarities and differences between the characteristics of, and the relationships among, the various social actors. Once again, I refer to the climate change regime matrix first presented in chapter 3 (table 3.5, p. 61). (For the sake of consistency; the matrix has since been updated to reflect each country's political status in 2003 and is reprinted here as table 7.1.)

Given that the ratification of an international treaty is the responsibility of national governments, it is impossible to understand any nation's position on the Kyoto Protocol without looking at the role of the state. Each of the three countries presented in this book provides cases of states with different levels of power: in Japan, the state is strong; in the United States, the state is weak; and in the Netherlands, the strength is intermediate. The characteristics of the nation-states discussed in this book are important to keep in mind, in that the strength of the state determines its level of autonomy in contrast to other social actors (see, e.g., Evans, Jacobson, and Putnam 1993). For example, as a result of the Japanese state's strength, the government has been able to shape its position on the Kyoto Protocol relatively unilaterally. Implementing measures to mitigate global climate change in Japan can therefore happen rather quickly. In contrast, the U.S. state is so fragmented that, throughout much of the nine years during which the United States was involved in the negotiations of a treaty to mitigate global climate change, the administrative and legislative branches of the government actively disagreed about what the U.S. position on the Kyoto Protocol should be. Although there was a brief period prior to the 2002 Senate elections when there was a Democrat-let Senate that began to push for reengagement in the international climate change negotiations,[3] the Bush administration refused to reconnect. In addition to the fragmentation of the U.S. government, members of the state have a very difficult time making decisions without the support of other social sectors, such as industry or civil society.

Like the state in each of these countries, the relationship of economic actors to other social actors, and to the overall political outcomes of the domestic position on the Kyoto Protocol, varies in each nation-state: in some cases, the relationship is collaborative, and in others, it is autonomous. As can be seen by the case studies in this book, the two counties that have relatively autonomous market sectors—the Netherlands and the United States—have had opposite responses to the Kyoto Protocol. Accordingly, the key factor may have less to do with the degree of autonomy of the market sectors than with its specific material composition, as discussed in chapter 6. The energy sector of the United States depends heavily on coal; thus, given the specific interests that derive from this fact, powerful economic actors have worked through conservative nonprofit organizations to limit U.S. involvement in the Kyoto Protocol (see, e.g., McCright and Dunlap 2000). Since rejecting the Kyoto Protocol, the Bush administration has pressured the U.S. government to be one of the few countries that will respond to energy needs by increasing levels of resource extraction rather than implementing efficiency standards and practices. In the Netherlands, as discussed in chapter 5, industry plays an altogether different role, leading the way in pushing to invest in the research and development of alternative technologies. Taking advantage of their connection to the Dutch state and their access to a relatively clean energy supply in the country's natural gas reserves, many types of industries in the Netherlands have moved to create an almost protectionist market for trying out alternative technologies while avoiding regulation.

Similar to the cases of those countries with autonomous market sectors, the two countries in which civil society has been external to the domestic decision-making process—Japan and the United States—have also had opposite responses to the protocol. In Japan, although civil society is external and locally based, the state, market, and science sectors have been working together to devise a progressive domestic climate change regime. It is likely that the political outcomes that we see in Japan are only possible as a result of the strong Japanese state that mediates the roles of and among other social actors. The United States is completely different. With its weak state and a civil society sector that is external to the climate change policy-making process, some members of the government have actually blamed the failure of the climate change regime on what they call "rabid environmentalists," or "rabid enviros" (interview, senior staff 2001). Others, however, say that without civil society's pressuring the U.S. government, climate change policy cannot happen. In contrast, civil society in the Netherlands has been internal to the policy-making process: citizens have voluntarily chosen to pay more for the electricity that powers their homes with a source that emits fewer greenhouse gases.

The role played by science in domestic climate change regime formation in these nation-states is less directive than that of the state, the market, or civil society. In fact, the cases of Japan, the Netherlands, and the United States present points on the range of science's engagement with the policy-making process—from relatively central to relatively peripheral. The most central of the cases—Japan—has seen the fewest questions regarding the validity of the science of climate change. In contrast to the Netherlands, where there is an imposed distance between scientists and policymakers, Japanese scientists frequently hold government positions, and they advise and socialize in the same social networks. It is probable that the centrality of the scientific community in Japan contributes to the fact that the science of the issue of climate change has been less challenged there than in either the Netherlands or the United States. In the United States, in particular, scientific uncertainty became a weapon that has been successfully wielded by conservative nonprofit organizations. Although the role that scientists are playing in the policy-making process is significant, the variable itself does not seem to be enough to explain the differences among these countries. In the words of Robert T. Watson, the former chair of the Intergovernmental Panel on Climate Change (IPCC), the centrality of science is a "necessary but not sufficient" precursor to the formation of an effective domestic climate change regime; it is "only one small input" (interview, July 23, 2001).

As can be seen by the depth of the interactions among various social actors within each nation-state, the relationships of each social actor in relation to the others are important. When seeking to understand each country's domestic position on the Kyoto Protocol, it is necessary to look at the interrelations among the state, market, civil society, and science in each country. As Habermas has stressed in his work, various social actors play central roles in maintaining legitimacy and control over advanced capitalist states (see, e.g., Habermas 1975). Throughout the years, Habermas has looked at the role of sociocultural variables outside of the state and the market, such as science and technology (Habermas 1970) and civil society (Habermas 1989, 1992, 1998; see also, Calhoun 1992). As suggested in Habermas's work, I have found that understanding the formation of a global climate change regime requires an appreciation of the roles of social actors within each nation and the role they play in affecting the global environmental system. Without looking at these actors and their interactions, it is impossible to understand the complexity of what is driving the formation of a state's domestic climate change regime and, thus, the international climate change regime itself.

Beyond these actors, another type of sociocultural variable plays a significant role in each country's decisions regarding climate change: the political culture of the country. Although Japan and the Netherlands are very different countries, they both have a tradition of collaboration—whether it be through the ancient notion of *wa* (harmony) or through the polder model. At the same time, certain aspects of American political culture, such as individualism, free enterprise, and private rights, have been capitalized on by conservative nonprofit organizations that are working to protect the interests of industries involved in the energy infrastructure in the United States.

THE GLOBAL ENVIRONMENTAL SYSTEM

Although it is necessary to look at the interrelations among domestic social actors to understand a country's position on the Kyoto Protocol, it is also important to conceptualize these relationships within a broader global environmental system; otherwise, important aspects of the global climate change regime will be missed. Analysis of the qualitative case studies in this book began by applying Ragin's Boolean-driven method of difference, to understand the patterns of various factors in the comparative cases (Ragin 1987). But to understand fully the political outcomes of each case, it is necessary to recognize that, although I have looked at each nation as a separate and comparable case, these countries are all part of an interactive global system. As has been stressed by McMichael (1990, 396), it is important to recognize the inherent interconnectedness of the countries of the world by developing social theories "through the comparative juxtaposition of elements of a dynamic, self-forming whole." In other words, the American, Japanese, and Dutch positions on the Kyoto Protocol have been decided, not just through the mediation of interrelated domestic actors, but also through each country's interactions with actors from other states and international organizations that were working on the protocol. Finally, each country's position on the Kyoto Protocol was determined by each nation's history and position within the global system in terms of its economic, political, and environmental characteristics.

It is likely that Japan, for example, only signed on to the agreement made in Bonn in response to international pressure from European and other countries, as well as the carbon sink concessions granted by the EU during the Bonn round of negotiations. In fact, by the end of the ministerial level negotiations, Japanese environment minister Kawaguchi announced, "Today's agreement is a vital step forward towards realizing the entry into force of the Kyoto Protocol by 2002" (Kawaguchi 2001b). As a nation that continues to be mired in an economic recession, the results of this meeting show that Japan is not in a position to isolate itself politically from its allies.

On the other side of the international coercion is the Netherlands, which has been attributed with successfully pressuring other members of the EU to push for more stringent emission reductions in the original text of the Kyoto Protocol in 1997 (Gummer and Moreland 2000). In Japan and the Netherlands, international actors worked in concert with domestic actors to affect the countries' decisions regarding their support of an international climate change regime.

In contrast, the United States was only able to remove itself from the climate change regime after almost nine years of negotiation because of its position in the global system as a political and economic leader. The United States, which has emerged as the unilateral global leader since the fall of the Soviet Union, in 1991, no longer seems particularly sensitive to global criticism. Even though many domestic actors did not support participation in the climate change negotiations during the Clinton administration, the administration seems to have tried very hard to avoid international criticism regarding its environmental policy. With the shift to the Bush administration, the United States began to pull out of a number of international regimes, including the Kyoto Protocol, expressing its ability to change its mind unilaterally—a luxury of the reigning global economic and political leader. Since its withdrawal from the Kyoto Protocol, the United States has gone on to resist international pressure on a number of occasions—with perhaps the most well known being its decision to go against UN consensus and attack Iraq, in March 2003. In fact, the actions taken by the United States since the September 11, 2001, attacks have been seen as the antithesis of multilateralism. In the words of Senator John Kerry, these actions represent "the too often belligerent and myopic unilateralism of the Bush Administration" (2003).

What Kerry is calling "myopic unilateralism" itself provides support for the global environmental system that I have presented within this book: it is only because of the global position of the United States in relation to the other countries of the world that America has been able to act in such a manner with regard to all types of international policy making. In other words, by looking at the global climate change regime through the lens of the global environmental system, I have provided an account of how domestic politics have serious global implications. Consistent with the arguments put forth in the work on the two-level game (Putnam 1988) and double-edged diplomacy (Evans, Jacobson, and Putnam 1993), what happens inside nation-states in response to international political issues is important. In contrast to the claims of many scholars working within the field of international relations, what happens inside nation-states is becoming increasingly more important rather than less.

ON POLITICAL AND MATERIAL OUTCOMES

Since the rules of compliance to the Kyoto Protocol were finalized in Marrakech in fall 2001 and domestic ratification began, it is possible to look more closely at the political outcomes to see if the countries were sincere when they said that they would move to ratify and bring the treaty to legal force by the end of 2002. Before we can question the economic and environmental implications of a global climate change regime that does not include the United States, however, we must first look toward bringing the treaty to legal force.

In terms of the United States, Japan, and the Netherlands, the positions put forth throughout this book have remained consistent. As has been previously stated, the Netherlands ratified the Kyoto Protocol on April 12, 2002, and, after much speculation about its intentions, Japan ratified the Kyoto Protocol on June 4, 2002. The United States, however, has continued to withdraw from international discussions about climate change. In fact, in summer 2003, when Senators McCain and Lieberman pushed for a vote on their proposed Climate Stewardship Act of 2003, debates about the validity of the science of climate change erupted on Capitol Hill once again. On July 28, 2002, Senator Inhofe—a senator from Oklahoma (a coal-producing state) and the chairman of the Senate Environment and Public Works Committee—gave a speech on the floor of the Senate regarding the political viability of the ratification of the Kyoto Protocol: the "95 Senators—both Democrats and Republicans—who, according to Byrd-Hagel, presumably oppose ratification if the [Kyoto] treaty came up on the Senate floor. . . . You have Senators who are of the liberal persuasion—fine people but certainly [of] a different philosophy than mine . . . who are really sincerely talking in favor of this Kyoto Treaty, but they cast their vote against it. They said: We don't want to ratify this treaty, and we are not going to ratify this treaty" (U.S. Senate, *Congressional Record,* 2003, S10015-6). In addition, the senator spoke at length about the lack of credibility of the science of climate change and the results of the Intergovernmental Panel on Climate Change (IPCC). In his own words, "Alarmists are attempting to enact an agenda of energy suppression that is inconsistent with American values, freedom, prosperity, and environmental problems. Over the past hour and a half I have offered compelling evidence that catastrophic global warming is a hoax" (U.S. Senate, *Congressional Record,* 2003, S10021). Although some senators tried to reopen discussion in the Senate about the regulation of greenhouse gases, the high-ranking chairman stated that discussion of the Kyoto Protocol is a waste of time, citing the unanimous vote on the Byrd-Hagel resolution in 1997.

Beyond the continuing debates within the United States, other countries have thus far blocked the protocol's entry into legal force. In particular, Australia and Russia have not ratified the Kyoto Protocol. Australia, which has been called the "world's biggest coal exporter,"[4] has stated that it will not consider ratification. Russia, however, originally reported that it would ratify the treaty by the end of 2002, but, at this point, it has also stated that "it agrees in principle to the treaty, but wants more time to study the plan."[5] During the climate change negotiations in December 2003, however, questions reemerged about whether Russia would ratify or not. Although newspapers around the world cited a Russian official's statement that Russia "shall not ratify [the Kyoto Protocol]" (Myers and Revkin 2003, A1), other sources reported a press conference with M. Tsikanov, the deputy minister of the economy and cochairman of the Russian Interagency Committee on Climate Change, who was quoted as stating that "Russia will ratify the protocol if it is proved that it is in our interest—all Russian political leaders have said so" (Guardian 2003). In other words, it is unclear if Russia will actually ratify the Kyoto Protocol. Without their ratification, however, the treaty will not have enough Annex I countries to meet the carbon dioxide requirements of 55 percent, as stated in article 25 of the Kyoto Protocol. In other words, even with the thirty industrialized countries that have ratified the protocol, they only account for 44.2 percent of the carbon dioxide emissions of developed countries; thus, the Kyoto Protocol cannot be brought to legal force.

Thus far, this concluding discussion has focused on political outcomes, the primary dependent variable of my book, while overlooking material outcomes. Given the purpose and nature of the climate change treaty, the material outcome of emission reductions is contingent on the success of the regime. In other words, the overall purpose of the Kyoto Protocol is to reduce greenhouse gas emissions. Since the text of the protocol and the rules for compliance to the regime were not completed until the end of the COP-7 negotiations, in fall 2001, it is hard to expect the regime to have been effective prior to its completion. At the same time, it would be expected that, if a country were sincere about meeting its emission reduction commitments, changes would already be visible. As has been pointed out throughout this book, national carbon dioxide emission levels in the countries that are the focus of this study have continued to rise since the Kyoto Protocol was drafted, in 1997.

Within the Bonn Agreement, many nation-states were allowed to account for emission reductions through forests that serve as carbon sinks, instead of enacting actual emission reductions. Some organizations and nation-states have been critical of the amount of carbon sinks that were allotted to countries such as Japan, Australia, Canada, and Russia when they threatened to pull out of the Kyoto Protocol. Without this provision, however, it is likely that the protocol would not have survived the COP-6bis round of the negotiations. First, this inclusion of sinks weakens the level of reductions—and thus the material outcome of the protocol—that will be seen in the first commitment period of 2008 to 2012. However, having a climate change regime that will motivate nations to invest in alternative energy sources and the development of new technologies prior to the second commitment period is likely to have positive effects on overall emissions during the commitment periods to come. Thus, although there will not be a significant decrease in greenhouse gas emissions during the first commitment period, which ends in 2012, the Kyoto Protocol has, in some ways, the potential to affect material outcomes successfully in the not-too-distant future.

At this point, it is impossible to predict the future of the Kyoto Protocol. In the best-case scenario, Russia will ratify the treaty as it has promised throughout the years of negotiations, and the protocol will become an environmental treaty that may provide the incentive for countries to move forward with alternative energy technologies. In the worst-case scenario, the treaty will never be brought to legal force, and 119 nations of the world will have wasted almost ten years negotiating an unratifiable treaty. Whatever the future, one thing is certain: without looking at national decision-making processes and their interactions within the global environmental system, it is impossible to understand international environmental policy making.

NOTES

1. Although a few European countries had begun the ratification process prior to the Marrakech Accords, most countries were unable to consider the ratification

of the Kyoto Protocol until the rules for compliance were finalized at the COP-7 meeting, in Marrakech.

2. United Nations Climate Change Secretariat, *Kyoto Protocol Status of Ratification*, 2003. Available at unfccc.int/resource/kpstats.pdf (accessed October 3, 2003).

3. As a result of Vermont senator Jim Jeffords' leaving the Republican Party on May 24, 2001, the Democrats became the majority in the Senate. However, the Republicans took back the majority during the midterm elections, in November 2002.

4. Reuters News Service, "World Biggest Coal Exporter Australia Dumps Kyoto," Canberra, Australia, June 6, 2002. Available at www.planetark.org/dailynewsstory.cfm/newsid/16298/story.htm (accessed October 3, 2003).

5. European Council for a Sustainable Energy Future, "Russia Wavers on Ratification of the Kyoto Protocol," September 29, 2003. Available at www.e5.org/modules.php?op=modload&name=News&file=article&sid=310 (accessed October 3, 2003).

Appendix A: People Interviewed in Japan

Mie Asaoka
President
Kiko Network
July 28, 1999

Yurika Ayukawa
Climate Change Campaign Officer
World Wide Fund (WWF) for Nature,
Japan
July 28, 1999

Ryuriko Demura
Member of the Board of Trustees
Seikatsu Club Consumers' Coopera-
tive Hokkaido, Green Fund
March 15–16, 2000

Akiko Domoto
Member of the House of Councillors
Japanese Diet
July 23, 1999 (written interview)

Hironori Hamanaka
Director General
Global Environment Department,
Environment Agency
July 20, 1999; November 16, 2000

Naoyuki Hata
Citizen's Forum
July 29, 1999

Taka Hiraishi
Cochair, Intergovernmental Panel on
Climate Change, National Green-house
Gas Inventories Program; and Senior
Consultant, Institute for Global Envi-
ronmental Strategies and Intergovern-
mental Panel on Climate Change
July 23, 1999; July 18, 2001; July 24, 2001

Kimiko Hirata
Kiko Network

July 29, 1999; April 3, 2000; November 15, 2000

Makoto Ikeuchi
Managing Director
Seikatsu Club Consumers' Cooperative, Hokkaido
March 16, 2000

Yasuko Kameyama
Researcher
National Institute of Environmental Studies
August 3, 1999

Kiyoshi Kawamura
Executive Director
Hokkaido Association of Towns and Villages
March 15, 2000

Yoichi Kaya
Director General
Research Institute of Innovative Technology for the Earth
August 2, 1999

Kojo Kunieda
Environmental and Nature Planning Office
Hokkaido Government
March 14, 2000

Kazuo Matsushita
Institute for Global Environmental Strategies
July 22, 1999

Nobu Mizushima
Hokkaido Prefecture New Energy Representative
March 15, 2000

Tsuneyuki Morita
Head of Environmental Economics Program

National Institute of Environmental Studies and Tokyo Institute of Technology
August 3, 1999

Masayo Nakase
Member of the Board of Trustees
Seikatsu Club Consumers' Cooperative, Hokkaido
March 16, 2000

Shuzo Nishioka
Graduate School of Media and Governance, Keio University; Executive Director, National Institute for Environmental Studies
July 22, 1999; November 21, 2000; July 22–23, 2001

Hajime Ohta
Executive Counselor
Keidanren-Japan Federation of Economic Organizations
August 4, 1999

Teruo Okazaki
Senior Manager
Global Environmental Affairs, Nippon Steel
July 27, 1999

Harumi Suda
Director
Citizen's Movement National Center
August 2, 1999

Toru Suzuki
Secretary General
Hokkaido Green Fund
March 15, 2000

Yasuyoshi Tanaka
Environment Writer
Mainichi Shimbun
March 9, 2000

Appendix B: People
Interviewed in the Netherlands

Hans Altevogt
Campaigner
Climate Change and Energy, Green-
peace Nederland
July 4, 2000

Magnus Andersson
Project Manager
COOL Europe, Wageningen University
August 8, 2000

Henriette C. Y. M. Bersee
Directorate General for Environmental
Protection
Ministry of Housing, Spatial Planning,
and the Environment
July 5, 2000

Merrilee Bonney
Directorate General for Environmental
Protection
Ministry of Housing, Spatial Planning,

and the Environment
June 30, 2000

Ilse Chang
Campaigner, Milieu Defensie (Friends
of the Earth)
November 16, 2000

Hans Davidse
Manager of Technology and Energy
Conservation
Akzo Nobel Energy
June 29, 2000

Joyeeta Gupta
Professor
Free University of Amsterdam
June 13, 2000

Wilhelma N. Kip
Executive Officer, European Affairs
EnergieNed—Federation of Energy

Companies in the Netherlands
June 9, 2000

Wiel Klerken
Director of Environmental Affairs
Confederation of Netherlands Industry and Employers (VNO-NCW)
June 14, 2000

Carolien Kroeze
Professor, Wageningen University
June 29, 2000

Peter W. Kwant
Group Research Advisor
Shell International B.V.
July 14, 2000

Bert Metz
Head of the Global Environmental Assessment Division
Dutch National Institute of Public Health and the Environment (RIVM)
July 12, 2000; November 28, 2000

Jeff Prins
Assistant to the Labor Party
Second Chamber of the Parliament
November 29, 2000

Marjolein Quené
Manager Business Development
NUON—International/Renewable Energy
June 16, 2000

Peter A. Scholten
Deputy Director General of Energy
Ministry of Economic Affairs
July 5, 2000

Sible Schöne
Manager, Climate Change Programme
World Wide Fund for Nature
June 7, 2000

Bart B. W. Thorborg
Ministry of Transport, Public Works, and Water Management
June 9, 2000

Andre van Amstel
Professor
Wageningen University
July 4, 2000

Marleen van de Kerkhof
Project Manager
COOL Nederland, Free University of Amsterdam
May 22, 2000

Gert N. van Ingen
President
Akzo Nobel Energy
June 29, 2000

Eimert van Middelkoop
Member of Dutch Parliament
Second Chamber
November 29, 2000

Pier Vellinga
Director
Institute for Environmental Studies, Free University of Amsterdam
June 13, 2000

J. Koos Verbeek
Coordinator, Climate Policy of the Ministry of Transport, Public Works, and Water Management
Royal Netherlands Meterological Institute
June 30, 2000

Robert T. Watson
Intergovernmental Panel on Climate Change
July 23, 2001

Appendix C: People Interviewed in the United States

Eileen Claussen
President
Pew Center on Global Climate Change
April 24, 2000

Congressional staff
Committee on Commerce
U.S. House of Representatives
August 11, 2000

Kert Davies
Science Policy Director
Ozone Action
January 19, 2000

David Doniger
Director of Climate Change Policy
Environmental Protection Agency
April 24, 2000

Nathalie Eddy
International Coordinator

U.S. Climate Action Network
April 27, 2000

William L. Fang
Deputy General Counsel
Edison Electric Institute
August 9, 2000

Foreign Relations Counsel
U.S. Senate
August 10, 2000

David Gardiner
Deputy Chairman
White House Climate Change Task Force
April 26, 2000

Bryan J. Hannegan
Staff Scientist
Committee on Energy and Natural Resources

U.S. Senate
May 14, 2001

Gerald Hapka
Deputy and General Counsel
Pew Center on Global Climate Change
April 24, 2000

Eric K. Holdsworth
Deputy Director
Global Climate Coalition
April 27, 2000

Glenn F. Kelly
Executive Director and CEO
Global Climate Coalition
April 27, 2000

Legislative assistant
U.S. Senate
May 11, 2001

Elizabeth L. Malone
Senior Research Scientist
Global Change Group
Pacific Northwest National Laboratory
April 27, 2000

Craig F. Montessano
Manager of Government Relations and
Public Affairs
Global Climate Coalition
April 27, 2000

Official A
U.S. Department of State
August 11, 2000

Official B
U.S. Department of State
August 11, 2000

Official C
U.S. Department of State
November 28, 2001

John W. Passacantando
Executive Director
Ozone Action
April 26, 2000

Chris Paynter
Executive Director
Greening Earth Society
August 9, 2000

Jonathan Pershing
Head of the Energy and Environment
Division
International Energy Agency
December 1, 2000; July 23, 2001

Hugh M. Pitcher
Staff Scientist
Global Change Group
Pacific Northwest National Laboratory
April 27, 2000

Rafe Pomerance
Chairman
Americans for Equitable Climate
Solutions
August 11, 2000; May 15, 2001

Dallas Scholes
Legislative Assistant and Counsel
Senator Michel B. Enzi
April 25, 2000

Senior official
Clinton Administration White
House
May 15, 2001

Senior staff
Committee on Energy and Natural Resources
U.S. Senate
May 11, 2001

Katherine A. Silverthorne
Climate Policy Officer
World Wildlife Fund
November 14, 2000

Steven J. Smith
Senior Research Scientist
Global Change Group, Pacific North-
west National Laboratory
April 27, 2000

Anthony Socci
Senior Climate Science Advisor
Office of Atmospheric Programs,
U.S. Environmental Protection
Agency
April 26, 2000; May 10, 2001

Edward R. Yawn
Director, Government Relations
Edison Electric Institute
August 9, 2000

Bibliography

Abramson, Paul R. "Postmaterialism and Environmentalism: A Comment on an Analysis and a Reappraisal." *Social Science Quarterly* 78, no. 1 (1997): 21–23.

Abramson, Paul R., and Ronald Inglehart. *Value Change in Global Perspective*. Ann Arbor: University of Michigan Press, 1995.

Adams, Richard M., Brian H. Hurd, Stephanie Lenhart, and Neil Leary. "Effects of Global Climate Change on World Agriculture: An Interpretive Review." *Climate Research* 11, no. 1 (1998): 19–30.

Anderson, Alison. *Media, Culture and the Environment*. New Brunswick, N.J.: Rutgers University Press, 1997.

Anderson, J. W. "Climate Change Diplomacy: The Next Step." *Resources* 142 (2001): 11–13.

Andersson, Magnus, and Joyeeta Gupta. "Addressing the Institutional Gaps in Global Environmental Governance." *Environment in the 21st Century* 1 (1998): 391–404.

Andresen, Steiner, Tora Skodvin, Arild Underdal, and Jorgen Wettestad. *Science and Politics in International Environmental Regimes*. Manchester: Manchester University Press, 2000.

Arnell, N. *Global Warming, River Flows, and Water Resources*, 99–149. New York: Wiley, 1996.

Ayres, Robert U. "Commentary on the Utility of the Ecological Footprint Concept." *Ecological Economics* 32 (March 2000): 347–49.

Baron, Richard. "Carbon and Energy Taxes in OECD Countries" (1997): 1–15. Available at www.iea.org/ieakyoto/docs/econinst/carboecd.pdf (accessed on January 9, 2004).

Barrett, Brendan, W. Bradnee Chambers, and Heike Schroeder. *Perceptions of Science and Politics in the UNFCCC Process: Delegates at COP3 and COP4*, 38. Tokyo: United Nations University, Institute of Advanced Studies, 2001.

Barrow, E. M., and M. Hulme. "Changing Probabilities of Daily Temperature Extremes in the U.K. Related to Future Global Warming and Changes in Climate Variability." *Climate Research* 6 (1996): 21–31.

Beck, Ulrich. "The Anthropological Shock: Chernobyl and the Contours of the Risk Society." *Berkeley Journal of Sociology* 32 (1987): 153–65.

——. *Ecological Politics in an Age of Risk*. Cambridge: Polity Press, 1995.

——. "Subpolitics." *Organization and Environment* 10 (1997): 52–65.

——. *World Risk Society*. Cambridge: Polity Press, 1999.

Beck, Ulrich, Anthony Giddens, and Scott Lash. *Reflexive Modernization: Politics, Tradition and Aesthetics in the Modern Social Order*. Stanford, Calif.: Stanford University Press, 1994.

Bell, Daniel. *The Coming of Post-industrial Society: A Venture into Social Forecasting*. New York: Basic Books, 1973.

Benedick, Richard. *Ozone Diplomacy*. Cambridge, Mass.: Harvard University Press, 1991.

Block, Fred. *Revising State Theory: Essays in Politics and Postindustrialism*. Philadelphia: Temple University Press, 1987.

Blowers, Andrew, and P. Glasbergen. *Prospects for Environmental Change*. New York: Wiley, 1996.

Blühdorn, Ingolfur. "Ecological Modernization and Post-ecologist Politics." In *Environment and Global Modernity*, edited by G. Spaargaren, A. P. J. Mol, and F. H. Buttel, 209–28. London: Sage Studies in International Sociology, 2000.

Bodansky, Daniel. "The History of the Global Climate Change Regime." In *International Relations and Global Climate Change*, edited by Urs Luterbacher and Detlef F. Sprinz, 23–40. Cambridge, Mass.: MIT Press, 2001.

Boehmer-Christiansen, S. "Global Climate Protection Policy: The Limits of Scientific Advice." *Global Environmental Change* 4 (1994): 140–59.

——. Climate Change: The World Bank, Global Environment Facility and the United Nations—Partners or Competitors in Global Environmental Governance." 1997: 1–45. Available at www.electromagnetism.demon.co.uk/z011.htm (accessed October 3, 2003).

Bord, R. J., A. Fisher, and R. E. O'Conner. "Public Perceptions of Global Warming: United States and International Perspectives." *Climate Research* 11 (1998): 74–85.

Bostrom, Ann, M. Granger-Morgan, Baruch Fischoff, and Daniel Read. "What Do People Know about Global Climate Change? 1. Mental Models." *Risk Analysis* 14, no. 6 (1994): 959–70.

Bradley, N. L., A. C. Leopold, J. Ross, and A. Huffaker. "Phenological Changes Reflect Climate Changes in Wisconsin." *Proceedings of the National Academy of Science* 96 (1999): 9701–4.

Brechin, Steven R. "Objective Problems, Subjective Values, and Global Environmentalism: Evaluating the Postmaterialist Argument and Challenging a New Explanation." *Social Science Quarterly* 80, no. 4 (1999): 793–809.

——. "Comparative Public Opinion and Knowledge on Global Climatic Change and the Kyoto Protocol: The U.S. versus the World?" *International Journal of Sociology and Social Policy* 23, no. 10 (2003): 106–34.

Brechin, Steven R., and Willett Kempton. "Global Environmentalism: A Challenge to the Postmaterialism Thesis?" *Social Science Quarterly* 75, no. 2 (1994): 245–69.

——. "Beyond Postmaterialist Values: National versus Individual Explanations of Global Environmentalism." *Social Science Quarterly* 78, no.1 (1997): 16–25.

Bunker, Stephen G. "Raw Material and the Global Economy: Oversights and Distortions in Industrial Ecology." *Society and Natural Resources* 9 (1996): 419–29.

Burniaux, Jean-Marc. "A Multi-gas Assessment of the Kyoto Protocol." OECD Economics Department Working Paper 270 (2000).

Buttel, Fredrick H. "World Society, the Nation-State, and Environmental Protection." *American Sociological Review* 65, no. 1 (2000a): 117–21.

——. "Classical Theory and Contemporary Environmental Sociology." In *Environment and Global Modernity*, edited by G. Spaargaren, A. P. J. Mol, and F. H. Buttel, 17–40. London: Sage Studies in International Sociology, 2000b.

——. "Ecological Modernization as Social Theory." *GeoForum* 31 (2000c): 57–65.

——. "Environmental Sociology and the Explanation of Environmental Reform." Paper presented at the Kyoto Environmental Sociology Conference. Kyoto: October 2001.

Calhoun, Craig, ed. *Habermas in the Public Sphere*. Cambridge, Mass.: MIT Press, 1992.

Cannell, M., and R. Smith. "Climatic Warming, Spring Budburst, and Frost Damage on Trees." *Journal of Applied Ecology* 20 (1986): 951–63.

Capella, Peter. "Global Warming Dangers 'Buried' in New Rankings." *Guardian* (London), January 29, 2001. Available at www.commondreams.org/views01/0129-03.htm (accessed January 9, 2004).

Carpenter, S. R., S. G. Fisher, N. B. Grimm, and J. F. Kitchell. "Global Change and Freshwater Ecosystems." *Annu. Rev. Ecol. Syst.* 23 (1992): 119–39.

Carter, K. K. "Provenance Tests as Indicators of Growth Response to Climate Change in 10 North Temperate Tree Species." *Canadian Journal for Forestry Research* 26 (1996): 1089–95.

Catling, Christopher, ed. *Insight Guides: Holland*. Boston: Houghton Mifflin, 1995.

Catton, William R., Jr. *Overshoot: The Ecological Basis of Revolutionary Change*. Urbana: University of Illinois Press, 1980.

Chambers, W. Bradnee, ed. *Global Climate Governance: Inter-linkages between the Kyoto Protocol and other Multilateral Regimes*. Tokyo: United Nations University, Institute for Advanced Studies, n.d.

Choucri, Nazli. *Global Accord: Environmental Challenges and International Responses*. Cambridge, Mass.: MIT Press, 1993.

Christoff, Peter. "Ecological Modernisation, Ecological Modernities." *Environmental Politics* 5 (1996): 476–500.

Climate Change Secretariat. *The Kyoto Protocol to the Convention on Climate Change*. Bonn, Germany: Climate Change Secretariat, 1998.

Clinton, William J., and Albert Gore Jr. *The Climate Change Action Plan*. Washington, D.C.: U.S. Department of State, 1993.

Cobb, Clifford, Ted Halstead, and Jonathan Rowe. "If the GDP Is Up, Why Is America Down?" *Atlantic Monthly* (October 1995). Available at www.theatlantic.com/politics/ecbig/gdp.htm accessed (October 3, 2003).

Cohen, Bonner R. "Russia Says It Will Not Ratify Kyoto Protocol." *Conference News Daily*, July 25, 2001, 1.

Cohen, Jean L., and Andrew Arato. *Civil Society and Political Theory*. Cambridge, Mass.: MIT Press, 1994.

Cohen, Maurie. "Science and the Environment: Assessing Cultural Capacity for Ecological Modernization." *Public Understanding of Science* 7 (1998): 149–67.

Cohen, Maurie J. "Ecological Modernization, Environment Knowledge, and National Character: A Preliminary Analysis of the Netherlands." In *Ecological Modernisation around the World: Perspectives and Critical Debates*, edited by A. P. J. Mol and D. A. Sonnenfeld, 77–107. Essex: Frank Cass, 2000.

Colijn, Helen. *Of Dutch Ways*. New York: Harper and Row, 1984.

Committee on the Human Dimensions of Climate Change, National Research Council. *Global Environmental Change: Understanding the Human Dimensions*. Washington, D.C.: National Academy Press, 1992.

———. *Human Dimensions of Global Environmental Change*. Washington, D.C.: National Academy Press, 1999.

Corliss, Mick. "Updated Environment Plan to Add New Economic Options." *Japan Times,* August 25, 2000, A9.

Council of the European Union. "Ten New Member States Set to Join the European Union." Press Release. 2003. Available at www.delcyp.cec.eu.int/en/news/16April/tenmembers.pdf (accessed October 2, 2003).

Cox, A., and N. O'Sullivan, eds. *The Corporatist State*. Aldershot, U. K.: Edward Elgar, 1988.

Crenshaw, E. M., and J. C. Jenkins. "Social Structure and Global Climate Change: Sociological Propositions Concerning the Greenhouse Effect." *Sociological Focus* 29 (1996): 341–58.

Cushman, John H., Jr. "U.S. Signs a Pact to Reduce Gases Tied to Warming." *New York Times,* November 13, 1998, A1.

Daly, Herman. "Sustainable Growth: An Impossibility Theorem." *Development* 3, no. 4 (1990): 45–47.

Daly, Herman, John B. Cobb Jr., and Clifford W. Cobb. *For the Common Good: Redirecting the Economy toward Community, the Environment, and a Sustainable Future*. Boston: Beacon Press, 1989.

Davis, Margaret B., and Ruth G. Shaw. "Range Shifts and Adaptive Responses to Quaternary Climate Change." *Science* 292 (2001): 673–79.

DeGroot, R. S., P. Ketner, and A. H. Ovaa. "Selection and Use of Bio-indicators to Assess the Possible Effects of Climate Change in Europe." *Journal of Biogeography* 22 (1995): 935–43.

Denzin, N. K., and Y. S. Lincoln. *Strategies of Qualitative Inquiry*. Thousand Oaks, Calif.: Sage Publications, 1998.

DeSombre, Elizabeth R. *Domestic Sources of International Environmental Policy: Industry, Environmentalists and US Power*. Cambridge, Mass.: MIT Press, 2000.

Dewey, John. *The Public and Its Problems*. New York: Holt, Rinehart and Winston, 1927.

Dietz, Thomas, and Eugene A. Rosa. "Effects of Population and Affluence on CO_2 Emissions." *Proceedings of the National Academy of Science* 94 (1997): 175–79.

Dobriansky, Paula J. "Remarks to Resumed Sixth Conference of Parties (COP-6) to the UN Framework Convention on Climate Change." Bonn, Germany. July 23, 2001. Available at www.state.gov/g/rls/rm/2001/4191pf.htm (accessed January 9, 2004).

Domhoff, G. William. *The Power Elite and the State: How Policy Is Made in America.* New York: A. de Gruyter, 1990.

Downs, A. "Up and Down with Ecology—The Issue Attention Cycle." *The Public Interest* 28 (1972): 38–50.

Downs, George W., Kyle Danish, and Peter N. Barsoom. "The Transformational Model of International Regime Design: Triumph of Hope or Experience?" *Columbia Journal of Transnational Law* 38, no. 3 (2000): 465–514.

Dryzek, John S. *The Politics of the Earth: Environmental Discourses.* Oxford: Oxford University Press, 1997.

Dunlap, Riley E. "Lay Perceptions of Global Risk: Public Views of Global Warming in Cross-National Context." *International Sociology* 13 (1998): 473–98.

Dunlap, Riley E., and Angela G. Mertig. *American Environmentalism: The U.S. Environmental Movement, 1970–1990.* Philadelphia: Taylor and Francis, 1992.

———. "Global Environmental Concern: An Anomaly for Postmaterialism." *Social Science Quarterly* 78, no. 1 (March 1997): 24–29.

Dunlap, Riley E., and Kent D. Van Liere. "The New Environmental Paradigm." *Journal of Environmental Education* 9 (Summer 1978): 10–19.

———. "Commitment to the Dominant Social Paradigm and Concern for Environmental Quality." *Social Science Quarterly* 65 (1984): 1013–28.

Dunn, Seth. "Can North and South Get in Step." *World Watch* 11, no. 6 (1998): 19–27.

Durkheim, Emile. *The Division of Labor in Society.* Translated by George Simpson. New York: Free Press, [1893] 1933.

Edelman, Murray. *The Symbolic Uses of Politics.* Chicago: University of Illinois Press, 1964.

Emirbayer, Mustafa, and Mimi Sheller. "Publics in History." *Theory and Society* 28 (1999): 145–97.

Energy Information Administration (EIA). *International Energy Annual 1999.* Washington, D.C.: Energy Information Administration, 1999. Available at www.eia.doe.gov/emeu/iea/tablee1.html (accessed October 3, 2003).

———. *United States of America-Energy Report,* 17. Washington, D.C.: Energy Information Administration, 2000. Available at www.eia.doe.gov/emeu/cabs/usa.html (accessed October 3, 2003).

Enevoldsen, Martin. "Industrial Energy Efficiency." *The Voluntary Approach to Environmental Policy,* edited by Arthur Mol, Volkmar Lauber, and Duncan Lieferink, 62–103. Oxford: Oxford University Press, 2000.

Environment Agency of Japan. "Law Concerning the Promotion of the Measures to Cope with Global Warming." Law number 117. July 1998.

Environmental Defense. "Environmental Defense Denounces Lost Opportunities at Climate Talks Stalemate." News release. November 25, 2000.

Environmental Protection Agency (EPA). Office of Policy, Planning, and Evaluation. *Climate Change and Wisconsin.* EPA 236-F-96-001. Washington, D.C., 1999.

European Foundation for the Improvement of Living and Working Conditions. "Wave of Reorganisations at Major Dutch Groups: An End to the Polder Model?" *European Industrial Relations Observatory Online,* 2003. Available at www.eiro.eurofound.ie/1998/11/features/NL9811106F.html (accessed October 2, 2003).

"EU Won't Alter View on Kyoto, Chirac Says." *Japan Times,* July 6, 2001.

Evans, Peter. *Embedded Autonomy.* Princeton, N.J.: Princeton University Press, 1995.

Evans, Peter B., Harold Karan Jacobson, and Robert D. Putnam. *Double-Edged Diplomacy: International Bargaining and Domestic Politics.* Berkeley: University of California Press, 1993.

Evans, Peter B., Dietrich Rueschemeyer, and Theda Skocpol. *Bringing the State Back In.* Cambridge: Cambridge University Press, 1985.

Falk, Richard. "Defining a Just War." *The Nation* 29 (October 2001): 11–15.

Farla, Jacco C. M., and Kornelis Blok. "Energy Efficiency and Structural Change in the Netherlands (1980–1995)." *Journal of Industrial Ecology* 4, no. 1 (2000): 93–117.

Fisher, Dana R. "Resource Dependency and Rural Poverty: Rural Areas in the United States and Japan." *Rural Sociology* 66 (2001): 181–202.

———. "From the Treadmill of Production to Ecological Modernization? Applying a Habermasian Framework to Society-Environment Relationships." *The Environmental State under Pressure* 10 (2002): 53–64.

Fisher, Dana R., and William R. Freudenburg. "Ecological Modernization and Its Critics: Assessing the Past and Looking toward the Future." *Society and Natural Resources* 14, no. 8 (2001): 701–9.

———. "Post-industrialization and Environmental Quality: An Empirical Analysis of the Environmental State." Presentation at the American Sociological Association Annual Meeting in Chicago, Illinois, August 2002.

Fisher, D. R., B. W. Hale, J. Hinke, and C. A. Overdevest. "Social and Ecological Responses to Climatic Change: Toward an Integrative Understanding." *International Journal of Environment and Pollution* 14, no. 4 (2002): 323–36.

Flavin, Christopher. "Last Tango in Buenos Aires." *World Watch* (November/December 1998): 11–18.

Forsberg, Aaron. *America and the Japanese Miracle.* Chapel Hill: University of North Carolina Press, 2000.

Foster, John Bellamy. "The Absolute General Law of Environmental Degradation under Capitalism." *Capitalism, Nature, Socialism* 2 (September 3, 1992): 77–82.

———. *Marx's Ecology: Materialism and Nature.* New York: Monthly Review Press, 2000.

Framework Convention on Climate Change. *Report of the Conference of the Parties on Its Third Session, Held at Kyoto from 1 to 11 December 1997.* United Nations. March 18, 1998.

Frank, David John. "Science, Nature, and the Globalization of the Environment 1870–1990." *Social Forces* 76 (1997): 409–35.

———. "The Social Bases of Environmental Treaty Ratification 1900–1990." *Sociological Inquiry* 69 (1999): 523–50.

Frank, David John, Ann Hironaka, and Evan Schofer. "Environmentalism as a Global Institution: Reply to Buttel." *American Sociological Review* 65, no. 1 (2000a): 122–27.

———. "The Nation-State and the Natural Environment over the Twentieth Century." *American Sociological Review* 65, no. 1 (2000b): 96–117.

Freeman, A. Myrick, III, and Robert H. Haveman. "Clean Rhetoric and Dirty Water." *The Public Interest* 28 (Summer 1972): 51–65.

Freudenburg, William R. "Addictive Economies: Extractive Industries and Vulnerable Localities in a Changing World Economy." *Rural Sociology* 57 (1992): 305–32.

———. "Poverty, Prosperity and Natural Resources in Wisconsin." Funding proposal. Madison: University of Wisconsin, 1998.

Freudenburg, William R., and Frederick H. Buttel. "Expert and Popular-Opinion Regarding Climate Change in the United States." In *Climate Change Policy in Germany and the United States*, 49–58. Berlin: Joseph Henry Press, 1997.

Freudenburg, William R., and Robert Gramling. "Natural Resources and Rural Poverty: A Closer Look." *Society and Natural Resources* 7 (1994): 5–22.

Friends of the Earth. "Hot Air! Two Weeks of Talk: US Ensures No Result." Media release. November 25, 2000.

Galanter, Marc. "Why the 'Haves' Come Out Ahead: Speculation on the Limits of Legal Change." *Law & Society Review* 9 (1974): 95–160.

Gallup Organization. "Americans Consider Global Warming Real, but Not Alarming." Gallup Poll News Service, 2001. Available at www.gallup.com/poll/releases/pr010409.asp (accessed October 3, 2003).

Garcia, C. A., S. Sabaté, and E. Tello. "Modeling the Responses to Climate Change of a Mediterranean Forest Managed at Different Thinning Intensities: Effects on Growth and Water Fluxes." *Impacts of Global Change on Tree Physiology and Forest Ecosystems*, edited by G. M. J. Mohren, K. Kramer, and S. Sabaté, 243–52. Norwell, Mass.: Kluwer Academic Publishers, 1997.

Gelbspan, Ross. *The Heat Is On*. Reading, Mass.: Addison-Wesley, 1997.

———. "Rx for a Planetary Fever." *The American Prospect*, May 8, 2000.

George C. Marshall Institute. "Climate Science and Policy: Making the Connection." Washington, D.C.: George C. Marshall Institute, 2001.

Giddens, Anthony. *The Third Way*. Cambridge: Polity Press, 1998.

Global Environmental Forum. *Environmental Data 1998*. Tokyo: Global Environmental Forum, 1998.

Global Leaders of Tomorrow Environment Task Force. *2001 Environmental Sustainability Index*. Report from the Annual Meeting of the World Economic Forum, Davos, Switzerland, 2001.

Global Warming Prevention Headquarters. "Guideline of Measures to Prevent Global Warming." June 19, 1998.

"Global Warming Update: Are Limits on Greenhouse Gas Emissions Needed?" *CQ Researcher* 6, no. 41 (November 1, 1996): 961–84.

Goldman, M. "Constructing an Environmental State: Eco-governmentality and Other Transnational Practices of a 'Green' World Bank." *Social Problems* 48 (2001): 499–523.

Goodell, Jeff. "How Coal Got Its Glow Back." *Conference News Daily*, reprinted from the *New York Times*, July 23, 2001, 6–9.

Gore, Albert, Jr. *Earth in the Balance*. Boston: Houghton Mifflin, 1992.

Gould, Kenneth A., Adam S. Weinberg, and Allan Schnaiberg. *Local Environmental Struggles: Citizen Activism in the Treadmill of Production*. New York: Cambridge University Press, 1996.

Gramling, Robert. *Oil on the Edge: Offshore Development, Conflict, Gridlock*. Albany: State University of New York Press, 1996.

Gramsci, Antonio. *Selections from the Prison Notebooks of Antonio Gramsci*. New York: International Publishers, 1971.

Grant, Wyn. *The Political Economy of Corporatism*. London: Macmillan, 1985.

"Greenhouse Gas Output Declines." *Japan Times*, September 23, 2000, A6.

Grimes, William W. *Unmaking the Japanese Miracle*. Ithaca: Cornell University Press, 2001.

Grubb, M. "The Skeptical Environmentalist—Measuring the Real State of the World." *Science* 294, no. 5545 (2001): 1285–87.

Gummer, John, and Robert Moreland. *The European Union and Global Climate Change: A Review of Five National Programmes*, 52. Washington, D.C.: Pew Center on Global Climate Change, 2000.

Guyette, R. P., and C. F. Rabeni. "Climate Responses among Growth Increments of Fish and Trees." *Oecologia* 104 (1995): 272–79.

Haas, Peter M. "Do Regimes Matter? Epistemic Communities and Mediterranean Pollution Control." *International Organization* 43 (1989): 377–403.

———. *Saving the Mediterranean: The Politics of Environmental Cooperation*. New York: Columbia University Press, 1990.

———. "Global Environmental Governance." In *Issues in Global Governance*, edited by Commission on Global Governance, 333–70. London, U. K.: Kluwer Law International, 1995.

———. "International Institutions and Social Learning in the Management of Global Environmental Risks." *Policy Studies Journal* 28, no. 3 (2000): 558–75.

Haas, Peter M., and Jan Sundgren. "Evolving International Environmental Law: Changing Practices of National Sovereignty." In *Global Accord: Environmental Challenges and International Responses*, edited by N. Choucri, 401–29. Cambridge, Mass.: MIT Press, 1993.

Hass, Peter M., Marc A. Levy, and T. Parson. "Appraising the Earth Summit: How Should We Judge UNCED's Success?" *Environment* 34, no. 8 (1992): 6–11, 26–33.

Habermas, Jürgen. "Technology and Science as 'Ideology.'" In *Towards a Rational Society*, chapter 6, 81–122. New York: Beacon, 1970.

———. *Legitimation Crisis*. Boston: Beacon Press, 1975.

———. *The Structural Transformation of the Public Sphere*. Cambridge, Mass.: MIT Press, 1989.

———. "Further Reflections on the Public Sphere." In *Habermas in the Public Sphere*, edited by Craig Calhoun, 421–61. Cambridge, Mass.: MIT Press, 1992.

———. *Between Facts and Norms*. Cambridge, Mass.: MIT Press, 1998.

Hajer, Maarten A. *The Politics of Environmental Discourse: Ecological Modernization and the Policy Process*. Oxford: Clarendon Press, 1995.

Hale, Cameron. "More Is Not Enough: National Energy and International Power." *Contemporary Journal of International Sociology* 34, no. 1 (1997): 17–38.

Hamilton, Lawrence C. *Modern Data Analysis: A First Course in Applied Statistics*. Pacific Grove, Calif.: Brooks/Cole, 1990.

Hann, Chris, and Elizabeth Dunn. *Civil Society: Challenging Western Models*. London: Routledge, 1996.

Hannigan, John. *Environmental Sociology: A Social Constructivist Perspective*. London: Routledge, 1995.

Hänninen, H. "Effects of Climatic Change on Overwintering of Forest Trees in Temperate and Boreal Zones." *Impacts of Global Change on Tree Physiology and*

Forest Ecosystems, edited by G. M. J. Mohren, K. Kramer, and S. Sabaté, 149–58. Norwell, Mass.: Kluwer Academic Publishers, 1997.

Harris, Paul G. *Understanding America's Climate Change Policy: Realpolitik, Pluralism, and Ethical Norms.* OCEES Research Paper Number 15. Oxford: Mansfield College, 1998.

Hawkins, Keith. *Environment and Enforcement: Regulation and the Social Definition of Pollution.* Oxford: Clarendon Press, 1984.

———. "Using Legal Discretion." In *A Reader on Administrative Law,* edited by D. J. Galligan, 247–73. London: Oxford University Press, 1996.

Heo, Jang. *Politics of Policy-Making: Environmental Policy Changes in Korea.* Ph.D. dissertation, University of Wisconsin, Madison, 1997.

Hideaki, Abe. "Environmental Standard Bearers." *Look Japan* (September 1999): 4–17.

Hogg, I. D., and D. D. Williams. "Response of Stream Invertebrates to a Global-Warming Thermal Regime: An Ecosystem-Level Manipulation." *Ecology* 77 (1996): 395–407.

Huber, J. *Die Regenbogengesellschaft. Ökologie und Sozialpolitik.* Frankfurt: Fisher Verlag, 1985.

———. *Unternehmen Umwelt. Weichenstellungen für Eine ökologische Marktwirtschart.* Frankfurt: Fisher, 1991.

Huber, Michael. "Leadership in the European Climate Policy: Innovative Policy Making in Policy Networks." In *The Innovation of EU Environmental Policy Initiatives,* edited by Duncan Liefferink and Mikael Skou Andersen, 133–55. Oslo: Scandinavian University Press, 1997.

Huddle, Norie, and Michael Reich. *Island of Dreams: Environmental Crisis in Japan.* Rochester, Vt.: Schenkman Books, 1975.

Idso, Craig D. *The Greening of Planet Earth: Its Progression from Hypothesis to Theory.* Arlington, Va.: Western Fuels Association, 1997.

Inglehart, Ronald. *Culture Shift in Advanced Industrial Society.* Princeton, N.J.: Princeton University Press, 1990.

———. "Public Support for Environmental Protection: Objective Problems and Subjective Values in 43 Societies." *PS: Political Science & Politics* 28, no. 1 (1995): 57–72.

Institute for Global Environmental Strategies (IGES). *Climate Policy Debate in Japan.* Report on the IGES Open Forum for Global Warming Abatement. November 2000. Tokyo: Institute for Global Environmental Strategies, 2000.

Intergovernmental Panel on Climate Change (IPCC). *The First Assessment Report of the Intergovernmental Panel on Climate Change.* Cambridge: Cambridge University Press, 1991.

———. *The Second Assessment Report of the Intergovernmental Panel on Climate Change.* Cambridge: Cambridge University Press, 1996.

———. Working Group I (IPCC WGI). *Climate Change 2001: The Scientific Basis—Third Assessment Report.* Cambridge: Cambridge University Press, 2001.

———. Working Group II (IPCC WGII). *Climate Change 2001: Impacts, Adaptation, and Vulnerability—Third Assessment Report.* Cambridge: Cambridge University Press, 2001.

———. Working Group III (IPCC WGIII). *Climate Change 2001: Mitigation—Third Assessment Report.* Cambridge: Cambridge University Press, 2001.

International Energy Agency (IEA). *Climate Change Policy Initiatives*. Paris: Organisation for Economic Co-operation and Development, 1992.

———. *Climate Change Policy Initiatives*. Paris: Organisation for Economic Co-operation and Development, 1994.

———. *Energy Balances of Organization for Economic Cooperation and Development Countries*. Paris: Organisation for Economic Co-operation and Development, 1999.

———. *The Road from Kyoto: Current CO_2 and Transport Policies in the IEA*. Paris: Organisation for Economic Co-operation and Development, 2000a.

———. *Dealing with Climate Change*. Paris: International Energy Agency and the Organisation for Economic Co-operation and Development, 2000b.

———. *Energy Statistics of OECD Countries, 1997–1998*. Paris: Organisation for Economic Co-operation and Development, 2000c.

———. *CO_2 Emissions from Fuel Combustion 1971–1998*. Paris: Organisation for Economic Co-operation and Development, 2001.

Jasanoff, Sheila, and Brian Wynne. "Science and Decisionmaking." In *Human Choice and Climate Change: The Societal Framework*, edited by S. Rayner and E. Malone, 1–88. Columbus, Ohio: Battelle Press, 1998.

Jones, P. D., T. J. Osburn, and K. R. Briffa. "The Evolution of Climate over the Last Millenium." *Science* 292 (2001): 662–67.

Josai Daigaku. *Review of Japanese Culture and Society*. Saitama-ken, Japan: Center for Intercultural Studies and Education, Josai University, 1986.

Joubert, A. M., and B. C. Hewitson. "Simulating Present and Future Climates of Southern Africa Using General Circulation Models." *Progress in Physical Geography* 21 (1997): 51–78.

Judis, John B. "Global Warming and the Big Shill." *The American Prospect,* February 1, 1999.

Kanie, Norichika. *Domestic Policy and Leadership in Multilateral Diplomacy: The Case of the Netherlands' Kyoto Protocol Negotiation, 28*. Tokyo: United Nations University Institute of Advanced Studies, n.d.

Kawaguchi, Yoriko. "Statement of the Minister for the Environment, Ms. Kawaguchi at the COP6 Resumed Session." July 19, 2001a. Bonn, Germany. Available at www.env.go.jp/en/topic/cc/010719.html (accessed January 9, 2004).

———. "Statement at the Final Ministerial Meeting of the COP6 Resumed Session." July 23, 2001b. Bonn, Germany. Available at www.env.go.jp/en/topic/cc/010723.html (accessed January 9, 2004).

Kawashima, Yasuko. "A Comparative Analysis of the Decision-Making Processes of Developed Countries toward CO_2 Emissions Reduction Targets." *International Environmental Affairs* 9 (1997): 95–126.

———. "Japan's Decision-Making about Climate Change Problems: Comparative Study of Decisions in 1990 and in 1997." *Environmental Economics and Policy Studies* 3 (2000): 29–57.

Keck, Margaret E., and Kathryn Sikkink. *Activists beyond Borders: Advocacy Networks in International Politics*. Ithaca, N.Y.: Cornell University Press, 1998.

Keohane, Robert, and Helen Milner. *Internationalization and Domestic Politics*. Cambridge: Cambridge University Press, 1996.

Kerry, John. "Foreign Policy Speech." Speech presented at Georgetown University, Washington, D.C., January 23, 2003.

Kidd, Quentin, and Aie-Rie Lee. "Postmaterialist Values and the Environment: A Critique and Reappraisal." *Social Science Quarterly* 78, no. 1 (1997a): 1–15.

———. "More on Postmaterialist Values and the Environment." *Social Science Quarterly* 78, no. 1 (1997b): 36–43.

Knight, John. "Making Citizens in Postwar Japan." In *Civil Society: Challenging Western Models*, edited by C. Hann and E. Dunn, 222–41. London: Routledge Press, 1996.

Kopp, R. J. *An Analysis of the Bonn Agreement*, 3. Washington, D.C.: Resources for the Future, 2001.

Krogman, Naomi. "Bureaucratic Slippage in Environmental Agencies: The Case of Wetlands Regulation." *Research in Social Problems and Public Policy* 7 (1999): 163–81.

Kull, Steven. "Americans on the Global Warming Treaty." *Program on International Policy Attitudes Online Report.* February 4, 2000. Available at www.pipa.org/OnlineReports/GlobalWarming/buenos_aires_02.00.html (accessed October 3, 2003).

Lambeck, Kurt, and John Chappell. "Sea Level Change through the Last Glacial Cycle." *Science* 292 (2001): 679–86.

Laumann, Edward O., and David Knoke. *The Organizational State.* Madison: University of Wisconsin Press, 1987.

Lee, Kai N. *Compass and Gyroscope: Integrating Science and Politics for the Environment.* Washington, D.C.: Island Press, 1993.

Leggett, Jeremy. *The Carbon War.* London: The Penguin Press, 1999.

Leroy, Pieter, and Jan van Tatenhove. "Political Modernization Theory and Environmental Politics." In *Environment and Global Modernity*, edited by G. Spaargaren, A. P. J. Mol, and F. H. Buttel, 187–208. London: Sage Studies in International Sociology, 2000.

Levy, David L., and Daniel Egan. "Capital Contests: National and Transnational Channels of Corporate Influence on the Climate Change Negotiations." *Politics and Society* 36 (1998): 337–62.

Levy, Marc A., Robert O. Keohane, and Peter M. Haas. "Improving the Effectiveness of International Environmental Institutions." In *Institutions for the Earth*, edited by R. O. Keohane, P. M. Haas, and M. A. Levy, 397–426. Cambridge, Mass.: MIT Press, 1993.

Lijphart, Arend. *The Politics of Accommodation: Pluralism and Democracy in the Netherlands.* Berkeley: University of California Press, 1975.

Lofland, John. *Social Movement Organizations: Guide to Research on Insurgent Realities.* New York: Aldine de Gruyter, 1996.

Lomborg, Bjørn. *The Skeptical Environmentalist: Measuring the Real State of the World.* Cambridge: Cambridge University Press, 2001.

Luterbacher, Urs, and Detlef F. Sprinz. *International Relations and Global Climate Change.* Cambridge, Mass.: MIT Press, 2001.

Lutzenhiser, Loren. "The Contours of U.S. Climate Non-Policy." *Society and Natural Resources* 14, no. 6 (2001): 511–24.

Magnuson, J. J., K. E. Webster, R. A. Assel, C. J. Bowser, P. J. Dillon, J. G. Eaton, H. E. Evans, and others. "Potential Effects of Climate Changes on Aquatic

Systems: Laurentian Great Lakes and Precambrian Shield Region." *Hydrological Processes* 11 (1997): 825–72.

Marx, Karl. *Capital: A Critique of Political Economy.* Translated by Ben Fowkes. New York: Vintage, 1997.

Mazur, A. "Global Environmental Change in the News: 1987–1990 versus 1992–1996." *International Sociology* 13, no. 4 (1998): 457–72.

Mazur, A., and J. Lee. "Sounding the Global Alarm: Environmental Issues in the U.S. National News." *Social Studies of Science* 23 (1993): 681–720.

McCarthy, John, and Meyer Zald. "Resource Mobilization and Social Movements: A Partial Theory." *American Journal of Sociology* 82, no. 6 (May 1977): 1212–41.

McCarthy, Thomas. *The Critical Theory of Jürgen Habermas.* Cambridge, Mass.: MIT Press, 1978.

McCright, Aaron M., and Riley E. Dunlap. "Challenging Global Warming as a Social Problem: An Analysis of the Conservative Movement's Counter-claims." *Social Problems* 47 (2000): 499–522.

McKean, Margaret. *Environmental Protest and Citizen Politics in Japan.* London: University of California Press, 1981.

McMichael, Philip. "Incorporating Comparison within a World-Historical Perspective: An Alternative Comparative Method." *American Sociological Review* 55 (1990): 385–97.

———. "Globalization: Myths and Realities." *Rural Sociology* 61 (1996): 25–55.

Meijnders, Anneloes. *Climate Change and Changing Attitudes.* Eindhoven, Neth.: Eindhoven University of Technology, 1998.

Mendelsohn, Robert O., and the American Enterprise Institute for Public Policy Research. *The Greening of Global Warming.* Washington, D.C.: AEI Press, 1999.

Messner, W., D. Bray, G. C. Germain, and N. Stehr. "Climate Change and Social Order: Knowledge for Action?" *Knowledge and Policy* 5 (1992): 82–100.

Metz, B., A. Faber, M. Berk, M. T. J. Kok, J. van Minnen, A de Moor. "From Kyoto to The Hague—European Perspectives on Making the Kyoto Protocol Work." The Bilt: RIVM, 2000.

Meyer, John W. "Rationalized Environments." In *Institutional Environments and Organizations,* edited by W. R. Scott, J. W. Meyer, and associates, 28–54. Thousand Oaks, Calif.: Sage, 1994.

Meyer, John W., John Boli, George M. Thomas, and Francisco O. Ramierez. "World Society and the Nation-State." *American Journal of Sociology* 103 (1997): 144–81.

Michaels, Patrick J. *Sound and Fury: The Science and Politics of Global Warming.* Washington, D.C.: Cato Institute, 1992.

Michaels, P. J., R. C. Balling Jr., R. S. Vose, and P. C. Knappenberger. "Analysis of Trends in the Variability of Daily and Monthly Historical Temperature Measurements." *Climate Research* 10 (1998): 27–33.

Miliband, Ralph. *The State in Capitalist Society.* New York: Basic, 1969.

Miller, Alan, and Mack McFarland. "World Responds to Climate Change and Ozone Loss." *Forum for Applied Research and Public Policy* 11, no. 2 (1996): 55–63.

Mills, C. Wright. *The Power Elite.* New York: Oxford University Press, 1959.

Mills, Mark P. *The Internet Begins with Coal: A Preliminary Exploration of the Impact of the Internet on Electricity Consumption.* Arlington, Va.: Greening Earth Society, 1999.

Mills, McCarthy, and Associates. *Does Price Matter? The Importance of Cheap Electricity for the Economy.* Arlington, Va.: Greening Earth Society, 1995.

Ministry of Housing, Spatial Planning, and the Environment. *The Netherlands' Climate Policy Implementation Plan: Part I.* The Hague: Ministry of Housing, Spatial Planning, and the Environment, 1999a.

———. *The Netherlands' Climate Policy Implementation Plan: Part II.* The Hague: Ministry of Housing, Spatial Planning, and the Environment, 1999b.

Mitsuda, Hisayoshi. "Surging Environmentalism in Japan: a Sociological Perspective." In *The International Handbook of Environmental Sociology,* edited by Redclift and Woodgate, 442–52. Cheltenham, U.K.: Edward Elgar, 1997.

Mitsuda, Hisayoshi, and Dana R. Fisher. "Environmental Sociology in Japan." *Environment & Society.* Newsletter of Research Committee 24 of the International Sociological Association. July 2000, 2–4.

Mol, Arthur P. J. *The Refinement of Production.* Utrecht, Neth.: Van Arkel, 1995.

———. "Ecological Modernization: Industrial Transformations and Environmental Reform." In *The International Handbook of Environmental Sociology,* edited by M. Redclift and G. Woodgate, 138–49. London: Edward Elgar, 1997.

———. "Ecological Modernization and the Environmental Transition of Europe: Between National Variations and Common Denominators." *Journal of Environmental Policy and Planning* 1 (1999): 167–81.

———. "The Environmental Movement in an Era of Ecological Modernisation." *Geoforum* 31 (2000a): 45–56.

———. "Globalization and Environment: Between Apocalypse-Blindness and Ecological Modernization." In *Environment and Global Modernity,* edited by G. Spaargaren, A. P. J. Mol, and F. H. Buttel, 121–50. London: Sage Studies in International Sociology, 2000b.

Mol, A. P. J., and D. A. Sonnenfeld, eds. *Ecological Modernisation around the World: Perspectives and Critical Debates.* Essex: Frank Cass, 2000.

Mol, Arthur P. J., and Gert Spaargaren. "Environment, Modernity and the Risk Society: The Apocalyptic Horizon of Environmental Reform." *International Sociology* 8 (1993): 431–59.

———. "Ecological Modernization Theory in Debate: A Review." In *Ecological Modernisation around the World: Perspectives and Critical Debates,* edited by A. P. J. Mol and D. A. Sonnenfeld, 17–49. Essex: Frank Cass, 2000.

Müller, Benito. "The Hague Climate Conference." December 2000. Available at www.wolfson.ox.ac.uk/~mueller (accessed October 3, 2003).

———. *The Resurrection of a Protocol: The Bonn Agreement and Its Impact on the "Climate Catch 22,"* 5. Oxford: Oxford Institute for Energy Studies, 2001.

Myers, Steven Lee, and Andrew C. Revkin. "Russia to Reject Pact on Climate, Putin Aide Says." *New York Times,* December 3, 2003, A1.

National Research Council. *Reconciling Observations of Global Temperature Change.* Washington, D.C.: National Academy Press, 2000.

———. *Climate Change Science: An Analysis of Some Key Questions.* Prepublication copy. Washington, D.C.: National Academy Press, 2001.

———. Committee on Alternative Energy Research and Development Strategies. *Confronting Climate Change: Strategies for Energy Research and Development.* Washington, D.C.: National Academy Press, 1990.

———. Committee on the Human Dimensions of Global Change. *Global Environmental Change: Understanding the Human Dimensions.* Washington, D.C.: National Academy Press, 1992.

New Hope Environmental Services. *State of the Climate Report: Essays on Global Climate Change.* Arlington, Va.: Greening Earth Society, 2000.

———. *In Defense of Carbon Dioxide: A Comprehensive Review of Carbon Dioxide's Effects on Human Health, Welfare, and the Environment.* Arlington, Va.: Greening Earth Society, n.d.

Nichols, David, and Eric Martinot. *Measuring Results from Climate Change Programs,* 50. Washington, D.C.: Global Environment Facility, 2000.

Nordlinger, Eric A. *On the Autonomy of the Democratic State.* Cambridge, Mass.: Harvard University Press, 1981.

NRC Handelsblad Webpagina. "The Political Branch of the Polder Model." *Profiel the Netherlands,* 1999. Available at www.nrc.nl/W2/Lab/Profiel/Netherlands/politics.html (accessed October 2, 2003).

O'Connor, James. "On the Two Contradictions of Capitalism." *Capitalism, Nature, Socialism* 2 (October 3, 1991): 107–9.

———. "Is Sustainable Capitalism Possible?" In *Is Capitalism Sustainable,* edited by James O'Connor, 152–75. New York: Guilford, 1994.

O'Connor, James R. *The Fiscal Crisis of the State.* New York: St. Martin's, 1973.

Offe, Claus. *Modernity and the State: East, West.* Cambridge, Mass.: MIT Press, 1996.

"Official: Russia May Sign Kyoto Protocol." *Guardian,* December 3, 2003. Available at www.guardian.co.uk/worldlatest/story/0,1280,-3459870,00.html (accessed January 9, 2004).

Ohta, Hajime. "Signs of Command and Control?" *Japan Times,* July 1998.

Organisation for Economic Co-operation and Development (OECD). *OECD Environmental Performance Review: Japan.* Paris: OECD, 1994.

———. *Climate Change: Mobilising Global Effort.* Paris: Organisation of Economic Co-operation and Development, 1997.

———. *Action against Climate Change: The Kyoto Protocol and Beyond.* Paris: Organisation for Economic Co-operation and Development, 1999a.

———. *National Climate Policies and the Kyoto Protocol.* Paris: Organisation for Economic Co-operation and Development, 1999b.

———. *OECD Environmental Data Compendium.* Paris: Organisation for Economic Co-operation and Development, 1999c.

O'Riordan, Tim, and James Cameron. *Interpreting the Precautionary Principle.* London: Earthscan Publications, 1996.

Ouchi, Kazuko. "Policy-Making after the Kyoto Conference." *Turning Up the Heat: Japanese NGO Responses to the Kyoto Conference and Japanese Climate Change Policy.* Edited and published by the Forum on Environmental Administration Reform (FEAR), Tokyo, 1998.

Parker, Larry B., and John E. Blodgett. *Global Climate Change Policy: From "No Regrets" to S. Res. 98,* 16. Washington, D.C.: Congressional Research Service, 1999. Available at www.ncseonline.org/NLE/CRSreports/Climate/clim-17.cfm?&CFID=10232082&CFTOKEN=97900275 (accessed October 3, 2003).

Paterson, Matthew. *Global Warming and Global Politics.* London: Routledge, 1996.

———. "Climate Policy as Accumulation Strategy: The Failure of COP6 and Emerging Trends in Climate Politics." *Global Environmental Politics* 1, no. 2 (2001): 10–17.

Paul, R., B. Dutta, S. Shattacharaya, A. P. Mitra, and M. Lal. "Kyoto Agreement on Greenhouse Gas Reduction and Future Global Temperature and Sea-Level Trends." *Current Science* 76 (1999): 1069–71.

Peterson, D. W., and D. L. Peterson. "Effects of Climate on Radial Growth of Sub-alpine Conifers in the North Cascade Mountains." *Canadian Journal for Forest Research* 24 (1994): 1921–32.

Pierce, John C. "The Hidden Layer of Political Culture: A Comment on 'Postmaterialist Values and the Environment: A Critique and Reappraisal.'" *Social Science Quarterly* 78, no. 1 (1997): 30–35.

Pimm, S., and J. Harvey. "The Skeptical Environmentalist: Measuring the Real State of the World." *Nature* 414, no. 6860 (2001): 149–50.

Porter, Gareth, and Janet Welsh Brown. *Global Environmental Politics.* Boulder, Colo.: Westview Press, 1991.

Poulantzas, Nicos. *State, Power, Socialism.* London: New Left Books, 1978.

Pronk, Jan. "Speech by Jan Pronk, Minister of Housing, Spatial Planning, and the Environment, to the Conference *Innovative Policy Solutions to Global Climate Change.*" Washington, D.C., April 25, 2000. Available at www.pewclimate .org/docUploads/jpronk%5Fspeech%2Epdf (accessed October 2, 2003).

———. "Statement at the Final Ministerial Meeting of the COP6 Resumed Session." Bonn, Germany, July 23, 2001.

Putnam, Robert D. "Diplomacy and Domestic Politics: The Logic of Two-Level Games." *International Organization* 42 (1988): 427–60.

Rabe, B. G. "The Politics of Global Climate Change: Implementing a 'Law of the Atmosphere' in American States and Canadian Provinces." *La Follette Policy Report* 10 (1999): 5–24.

Ragin, Charles. *The Comparative Method: Moving beyond Qualitative and Quantitative Strategies.* Berkeley: University of California Press, 1987.

Rayner, Steve. "Global Environmental Change: Understanding the Human Dimensions of Global Change." *Environment* 34, no. 7 (1992): 25–28.

Rayner, Steve, and Elizabeth Malone. *Human Choice and Climate Change.* Columbus, Ohio: Battelle Press, 1998.

Read, D., A. Bostrom, and M. Granger-Morgan. "What Do People Know about Global Climate Change?" *Risk Analysis* 14 (1994): 971–82.

Repetto, Robert. *Jobs, Competitiveness, and Environmental Regulation: What Are the Real Issues?* Washington, D.C.: World Resources Institute, 1995.

Revkin, Andrew C. "When Will We Be Sure?" *New York Times,* September 10, 2000, WK3.

Risse-Kappen, Thomas, ed. *Bringing Transnational Relations Back In: Non-state Actors, Domestic Structure and International Institutions.* Cambridge: Cambridge University Press, 1995.

Roberts, J. Timmons. "Global Inequality and Climate Change." *Society and Natural Resources* 14, no. 6 (2001): 501–24.

Roberts, J. Timmons, and Peter E. Grimes. "Carbon Intensity and Economic Development 1962–1991." *World Development* 25, no. 2 (1997): 191–98.

Rosa, Eugene A. "Global Climate Change: Background and Sociological Contributions." *Society and Natural Resources* 14, no. 6 (2001): 491–500.

Rosa, Eugene A., and Thomas Dietz. "Climate Change and Society: Speculation, Construction and Scientific Investigation." *International Sociology* 13, no. 4 (1998): 421–55.

Rowlands, Ian H., and Malory Greene. *Global Environmental Change and International Relations.* Basingstoke, Hampshire, U. K.: Macmillan Academic and Professional, 1992.

Rudel, Thomas K. "Sequestering Carbon in Tropical Forests: Experiments, Policy Implications, and Climatic Change." *Society and Natural Resources* 14, no. 6 (2001): 525–32.

Sarewitz, Daniel, and Roger Pielke Jr. "Breaking the Global Warming Gridlock." *The Atlantic Monthly,* July 2000. Available at www.theatlantic.com/issues/2000/07/.htm (accessed January 5, 2004).

Schindler, D. W. "Widespread Effects of Climatic Warming on Freshwater Ecosystems in North America." *Hydrological Processes* 11 (1997): 1043–67.

Schnaiberg, Allan. *The Environment: From Surplus to Scarcity.* New York: Oxford University Press, 1980.

Schnaiberg, Allan, and Kenneth Gould. *Environment and Society: The Enduring Conflict.* New York: St. Martin's Press, 1994.

Schneider, S. "Global Warming: Neglecting the Complexities." *Scientific American* 286, no. 1 (2002): 62–65.

Schramm, Gunter. "Electric Power in Developing Countries: Status, Problems, Prospects." In *Annual Review of Energy 1990.* Palo Alto, Calif.: Annual Reviews, 1990.

Schreurs, M. A., and E. C. Economy, eds. *The Internationalization of Environmental Protection.* Cambridge: Cambridge University Press, 1997.

Schreurs, Miranda. 1997. "Domestic Institutions and International Environmental Agendas in Japan and Germany." In *The Internationalization of Environmental Protection,* edited by M. A. Schreurs and E. C. Economy, 134–61. Cambridge: Cambridge University Press.

Schroeder, Heike. *Negotiating the Kyoto Protocol: An Analysis of Negotiation Dynamics in International Negotiations.* Munster: LIT Publisher, 2001.

Schwartz, M. D. "Monitoring Global Change with Phenology: The Case of the Spring Green Wave." *International Journal of Biometeorology* 38 (1994): 18–22.

Shanahan, J., and C. Trumbo. *Voices and Messages in Global Climate Change Coverage.* Unpublished manuscript, 1998.

Silverman., D. *Qualitative Research: Theory, Method and Practice.* Thousand Oaks, Calif.: Sage Publications, 1997.

Singer, S. F. *Hot Talk Cold Science: Global Warming's Unfinished Debate.* Oakland, Calif.: The Independent Institute, 1998.

Skocpol, Theda. *States and Social Revolutions: A Comparative Analysis of France, Russia, and China.* Cambridge: Cambridge University Press, 1979.

———. "Bringing the State Back In: Strategies of Analysis in Current Research." In *Bringing the State Back In,* edited by Peter R. Evans, Dietrich Ruesdchemeyer, and Theda Skocpol, 3–37. Cambridge: Cambridge University Press, 1985.

Skodvin, Tora. "The Intergovernmental Panel on Climate Change." In *Science and Politics in International Environmental Regimes,* edited by Steiner Andresen, Tora

Skodvin, Arild Underdal, and Jorgen Wettestad, chapter 7, 146–80. Manchester: Manchester University Press, 2000.

Smith, Jackie, Charles Chatfield, and Ron Pagnucco, ed. *Transnational Social Movements and Global Politics: Solidarity beyond the State.* Syracuse, N.Y.: Syracuse University Press, 1997.

Smith, Robert John. *Japanese Society: Tradition, Self, and the Social Order.* Cambridge: Cambridge University Press, 1983.

Soroos, Marvin S. *The Endangered Atmosphere: Preserving a Global Commons.* Columbia: University of South Carolina Press, 1997.

———. "The Thin Blue Line: Preserving the Atmosphere as a Global Commons." *Environment* 40, no. 2 (1998): 7–13, 32–35.

———. "Global Climate Change and the Futility of the Kyoto Process." *Global Environmental Politics* 1, no. 2 (2001): 1–9.

Spaargaren, Gert. *The Ecological Modernization of Production and Consumption: Essays in Environmental Sociology.* Wageningen, Neth.: Wageningen University, 1997.

———. "Ecological Modernization Theory and the Changing Discourse on Environment and Modernity." In *Environment and Global Modernity*, edited by G. Spaargaren, A. P. J. Mol, and F. H. Buttel, 41–72. London: Sage Studies in International Sociology, 2000.

Spaargaren, Gert, and Arthur P. J. Mol. "Sociology, Environment, and Modernity: Ecological Modernization as a Theory of Social Change." *Society and Natural Resources* 5 (1992): 323–44.

Spaargaren, Gert, Arthur P. J. Mol, and Fredrick H. Buttel. *Environment and Global Modernity.* London: Sage Studies in International Sociology, 2000.

Spaargaren, Gert, and Bas van Vliet. "Lifestyles, Consumption and the Environment: The Ecological Modernisation of Domestic Consumption." In *Ecological Modernisation around the World: Perspectives and Critical Debates*, edited by A. P. J. Mol and D. A. Sonnenfeld, 50–76. Essex, U. K.: Frank Cass, 2000.

Sparks, T. H., and P. D. Carey. "The Responses of Species to Climate over Two Centuries: An Analysis of the Marsham Phenological Record, 1736–1947." *Journal of Ecology* 83 (1995): 321–29.

Sprinz, Detlef F., and Martin Weiss. "Domestic Politics and Global Climate Policy." In *International Relations and Global Climate Change*, edited by Urs Luterbacher and Detlef F. Sprinz, 67–94. Cambridge, Mass: MIT Press, 2001.

Stearns, Lisa. "Fact and Fiction of a Model Enforcement Bureaucracy: The Labor of Inspectors of Sweden." *British Journal of Law and Society* 6 (1979): 1–23.

Stigler, George J. *The Citizen and the State: Essays on Regulation.* Chicago: University of Chicago Press, 1975.

Stokke, O. S., and Ø. B. Thomessen, eds. *Yearbook of International Co-operation on Environment and Development.* London: Earthscan, 2001.

Stone, C. D. "Beyond Rio: 'Insuring' against Global Warming." *American Journal of International Law* 86 (1992): 445–88.

Stone, Peter Bennet. *Japan Surges Ahead: The Story of an Economic Miracle.* New York: Praeger, 1969.

Sugiyama, Taishi, and Axel Michaelowa. "What Must and Can COP6 Decide? Extended Flexibility Backed by Transparency and Responsibility." *Energy Policy* 28 (2000): 571–74.

Templett, Paul H., and Stephen Farber. "The Complementarity between Environment and Economic Risk: An Empirical Analysis." *Ecological Economics* 9 (1994): 153–65.

Thompson, Michael, and Steve Rayner. "Cultural Discourse." In *Human Choice and Climate Change*, vol. 1, edited by S. Rayner and E. L. Malone, 265–344. Columbus, Ohio: Battelle Press, 1998.

"Time/CNN Poll: We Want Action—Unless the Price Is Too High." *Time Magazine*, April 9, 2001, 32.

Tolba, Mostafa Kamal, and Iwona Rummel-Bulska. *Global Environmental Diplomacy: Negotiating Environmental Agreements for the World, 1973–1992.* Cambridge, Mass.: MIT Press, 1998.

Tonnies, Ferdinand. *Fundamental Concepts in Sociology (Gemeinschaft und Gesellschaft).* Translated by Charles P. Loomis. New York: American Book, [1887] 1963.

"Toxic Waste in Japan: The Burning Issue." *The Economist,* July 25, 1998, 60.

Trilling, Julia, and Steinar Strom, eds. *Global Climate Change: European and American Policy Responses.* Berkeley: Center for Western European Studies, University of California Berkeley, 1993.

Trumbo, C. "Constructing Climate Change: Claims and Frames in U.S. News Coverage of an Environmental Issue." *Public Understanding of Science* 5 (1996): 269–83.

Ungar, Sheldon. "Bringing the Issue Back In: Comparing the Marketability of the Ozone Hole and Global Warming." *Social Problems* 45, no. 4 (1998): 510–27.

Union of International Associations (UIA). *Yearbook of International Organizations.* Munich, Germany: K.G. Saur, 2000.

United Nations Climate Change Secretariat. *Framework Convention on Climate Change.* France: United Nations Environmental Programme, 1992.

———. *Kyoto Protocol to the Convention on Climate Change.* France: UNEP, 1998.

———. *Report of the Conference of the Parties on Its First Session, Held at Berlin from 28 March to 7 April 1995—Addendum.* Bonn, Germany, 1995.

———. *Report of the Conference of the Parties on Its Second Session, Held at Geneva from 8 to 19 July 1996—Addendum,* 1996.

United Nations University. *Global Climate Governance: Inter-linkages between the Kyoto Protocol and Other Multilateral Regimes: Final Report.* Tokyo: United Nations University, Institute of Advanced Study, Global Environmental Information Centre, 1999.

U.S. Department of State. *National Action Plan for Global Climate Change.* Washington, D.C.: U.S. Department of State, 1992.

U.S. Senate. "Senate Resolution 98." *Congressional Record.* Report no. 105-54. June 12, 1997.

———. "Proceedings and Debates of the 105th Congress, Expressing Sense of Senate Regarding U.N. Framework Convention on Climate Change." *Congressional Record.* S8113-S8139. July 25, 1997.

———. "Science of Climate Change." *Congressional Record 2003.* S10012-23. July 28, 2003.

van Tatenhove, Jan, Bas Arts, and Pieter Leroy. *Political Modernisation and the Environment.* Dordrecht: Kluwer Academic Publishers, 2000.

van Wolfren, Karel. *The Enigma of Japanese Power.* New York: A. A. Knopf, 1989.

Vellinga, Pier, and Michael Grubb, eds. *Climate Change Policy in the European Community.* London: Energy and Environmental Program, Royal Institute of International Affairs, 1992.

Victor, David. *The Collapse of the Kyoto Protocol and the Struggle to Slow Global Warming.* Princeton, N.J.: Princeton University Press, 2001.

Vig, Norman J., and Regina S. Axelrod. *The Global Environment: Institutions, Law, and Policy.* Washington, D.C.: CQ Press, 1999.

von Storch, Hans, and Dennis Bray. "Perspectives of Climate Scientists on Global Climate Change." In *Climate Change Policy in Germany and the United States,* 33–48. Berlin: German-American Academic Council Foundation, 1997.

Vrolijk, Christiaan. "COP-6 Collapse or 'To Be Continued . . .'." *International Affairs* 77, no. 1 (2001): 163–69.

Wackernagel, Mathis, Larry Onisto, Alejandro Callejas Linares, Ina Susana Lopez Falfan, Jesus Mendez Garcia, Ana Isabel Suarez Guerrero, Ma. Guadalupe Suarez Guerrero. *Ecological Footprint of Nations.* Xalapa, Mexico: Centro de Estudios para la Sustentabilidad, Universidad, Anahuac de Xalapa, 1997.

Wackernagel, Mathis, and William E. Rees. *Our Ecological Footprint: Reducing Human Impact on the Earth.* Gabriola Island, B.C.: New Society Publishers, 1996.

Wallerström, Margot. European Union press conference. Bonn, Germany. July 20, 2001.

Wapner, Paul Kevin. *Environmental Activism and World Civic Politics.* Albany: State University of New York Press, 1996.

Watson, Robert T. "Presentation of the Chair of the Intergovernmental Panel on Climate Change at the Sixth Conference of Parties to the United Nations Framework Convention on Climate Change." The Hague, the Netherlands. November 13, 2000. Transcript available at www.ipcc.ch/press/sp-cop6.htm (accessed January 9, 2004).

Weber, Max. *The Protestant Ethic and the Spirit of Capitalism.* Translated by Talcott Parsons. New York: Scribner's Sons, 1958.

Weingart, Peter, Anita Engels, and Petra Pansegrau. "Risks of Communication: Discourses on Climate Change in Science, Politics, and the Mass Media." *Public Understanding of Science* 9 (2000): 261–83.

Wilson, Edward O. "The Ecological Footprint." *Vital Speeches of the Day* 67, no. 9 (February 15, 2001): 274–78.

Winnet, S. M. "Potential Effects of Climate Change on U.S. Forests: A Review." *Climate Research* 11 (1998): 39–49.

World Resources Institute. *World Resources 1994–1995.* New York: Oxford University Press, 1994.

Wright, Scott D., and Dale A. Lund. "Gray and Green? Stewardship and Sustainability in an Aging Society." *Journal of Aging Studies* 14, no. 3 (September 2000): 229–50.

Wuthnow, Robert. *Between States and Markets: The Voluntary Sector in Comparative Perspective.* Princeton, N.J.: Princeton University Press, 1991.

Wynne, Brian. "Scientific Knowledge and the Global Environment." In *Social Theory and the Global Environment,* edited by Redclift and Benton, 169–89. London: Routledge, 1994.

Yeager, Peter C. *The Limits of the Law: The Public Regulation of Private Pollution.* New York: Cambridge University Press, 1990.

York, Richard, Eugene A. Rosa, and Thomas Dietz. "Footprints on the Earth: The Environmental Consequences of Modernity." *American Sociological Review* 68, no. 2 (April 2003): 279–300.

Young, Oran. "The Politics of International Regime Formation: Managing Natural Resources and the Environment." *International Organization* 43 (1989): 340–75.

———. *International Governance: Protecting the Environment in a Stateless Society.* Ithaca, N.Y.: Cornell University Press, 1994.

———. *Creating Regimes: Arctic Accords and International Governance.* Ithaca, N.Y.: Cornell University Press, 1998.

———, ed. *Global Governance: Drawing Insights from the Environmental Experience.* Cambridge, Mass.: MIT Press, 1997.

Young, Oran R., George J. Demko, and Kilaparti Ramakrishna. *Global Environmental Change and International Governance.* Hanover, N.H.: University Press of New England (for Dartmouth College), 1996.

Zachos, James, Mark Pagani, Lisa Sloan, Ellen Thomas, and Katharina Billups. "Trends, Rhythms, and Aberrations in Global Climate 65 Ma to Present." *Science* 292 (2001): 686–93.

Zald, Meyer N., and John D. McCarthy. "America and the Rise of Social Movements." *New Society,* June 28, 1972, 670–72.

Index

About the Author

Dana R. Fisher is assistant professor in the Department of Sociology at Columbia University. She has published numerous articles on the politics of climate change and political decision making.